An Introduction to Signal Processing for Non-Engineers

T0134064

An Introduction to
Signal Processing for
Non-Engineers

Afshin Samani

CRC Press
Taylor & Francis Group
Boca Raton London New York

CRC Press is an imprint of the
Taylor & Francis Group, an **informa** business

CRC Press
Taylor & Francis Group
6000 Broken Sound Parkway NW, Suite 300
Boca Raton, FL 33487–2742

First issued in paperback 2022

© 2020 by Taylor & Francis Group, LLC

CRC Press is an imprint of Taylor & Francis Group, an Informa business

No claim to original U.S. Government works

ISBN-13: 978-0-367-20755-7 (hbk)
ISBN-13: 978-1-03-233781-4 (pbk)
DOI: 10.1201/9780429263330

Library of Congress Cataloging-in-Publication Data

Names: Samani, Afshin, author.
Title: An introduction to signal processing for non-engineers / by Afshin Samani.
Description: First edition. I Boca Raton, FL : CRC Press/Taylor & Francis Group, 2019. I Includes bibliographical references and index.
Identifiers: LCCN 2019029229 (print) I LCCN 2019029230 (ebook) I ISBN 9780367207557 (hardback) I ISBN 9780429263330 (ebook)
Subjects: LCSH: Signal processing.
Classification: LCC TK5102.9 .S265 2019 (print) I LCC TK5102.9 (ebook) I DDC 621.382/2—dc23
LC record available at https://lccn.loc.gov/2019029229
LC ebook record available at https://lccn.loc.gov/2019029230

Visit the Taylor & Francis Web site at www.taylorandfrancis.com

and the CRC Press Web site at www.crcpress.com

eResource material is available for this title at https://www.crcpress.com/9780367207557.

Contents

Preface

I have a degree in biomedical engineering, but when I started my career as a teacher at the Department of Health, Science and Technology in Aalborg University, I came across many students who did not have an engineering or mathematical background and therefore had difficulty following the basic principles of signal processing. Despite this drawback, they had to go through all the steps of measuring biomechanical or biological signals in the lab and interpreting their results. Later, I occasionally came across researchers and PhD students in the field who still did not have a full comprehension of the basic principles of signal processing.

I was not the only person at our department who was faced with this challenge. Therefore, my colleagues established a new line of education for post-graduate students in sport science entitled "sport technology" aimed at covering this gap in knowledge and educating students so they are not only familiar with their respective field of sport science but also have an acceptable understanding of basic technical concepts and methods in their fields. The aim was to have students graduate who are capable of performing technical projects and also working at companies and create a bridge between the end users and developers of new technologies.

I am also part of the team that tries to fulfill this educational aim. My assignment was to teach basic principles of measurement and signal processing. I realized that I cannot simply adopt typical textbooks in this field and rely on students to follow all the introduced concepts and methods. The students did not have the mathematical background to fully comprehend those textbooks, and it was not within the scope of the course to dig that deep into the field either. Therefore, I had to find a mediating language and style to deliver the message but at the same time not scare the students with the complexity of the topic. It took one or two semesters until I felt comfortable with the approach I adopted to teach in that course.

I was rather happy, and students seemed to follow the course without major complaints. After a few semesters, though, I had a complaint from one of the students who was commenting about the lack of sources to prepare them for the exam. This was quite surprising to me, because I had already introduced them to a book that was rather light in terms of mathematical complexity, and this was in addition to the notes, slides and exercises that we had during the course. The book did not cover everything I taught them in the course, but I was hoping that a combination of their own notes, slides and the book would be sufficient for them to prepare for the exam. I was apparently incorrect.

Finding a book with a low degree of mathematical complexity and the same extent of coverage that I used to teach in the course was not an easy task. This made me think of my role as a teacher: What can I do to help them further? At first, I thought maybe I had not thoroughly searched for a relevant textbook for this course or maybe there is a new book that I had not the chance to see until then. I gave it another shot, but I was right in the sense that the available books are either quite complex in terms of math, quite bulky or may be discouraging for a beginner to start reading, as they were merely theoretical books with very few practical examples.

I thought of writing a book to fill this gap and provide a reference book introducing the basic concepts of signal processing for this targeted group of students
(and perhaps scientists) with minimum use of mathematical formulations and more
emphasis on visual illustrations. I talked with a few colleagues, and they encouraged
me to go down that path. They also had the same experience that many students, and
even occasionally researchers, do not have a basic understanding of the methods,
which compromises their ability to interpret experimental results.

Thus, I aimed at presenting an intuitive approach to understand the basics of signal processing such that readers achieve basic knowledge and skills in this topic. In
my view, this idea would work if I keep the verbosity of the book to a minimum and
let figures talk more than words. Wherever relevant, I also included MATLAB files
to generate the figures presented in the book. Thus, students get to know some tricks
in MATLAB to make their own figures. This book should not ever be regarded as a
thorough reference book of signal processing, and no mathematical proofs for any of
the theorems should be expected here. However, after reading this book, the reader
should have an informed view about the applied methods and interpretation of the
results. This may even encourage them to dig further in the field and deal with the
existing complexities.

Acknowledgments

From the start, I had to take on this book project as a side activity in addition to my daily job duties and my personal life. This book here is the result of spending hours and hours of my spare time on the weekends and extra hours after the working day. I was involved in every part of this book, from A to Z, and did not have much assistance. However, there are still a few people and entities that I would like to acknowledge. I would really like to thank CRC Press for seeing a potential in this book. I would like to thank the students who were my focus when I thought about initiating this project. I would like to acknowledge Aalborg University for providing an exciting working environment, which was an important energy source when I wanted to embark on such a journey. Further, my good colleagues who contribute to create a constructive and inspiring atmosphere at work. I owe a debt to all of my teachers during my own schooling for everything I learned about signal processing. I hope I have been successful in passing on this knowledge to the next generation. I hope that the students who read this book find it useful to achieve a good understanding of the basic concepts in signal processing. Finally, yet importantly, I would very much like to thank my family and my beloved wife, Shiva, for her kindness, patience and support. I am sure that her kind heart excuses me for all the time I left her alone to work on this project.

Author

Afshin Samani earned a PhD in biomedical engineering and science in 2010 at Aalborg University, Denmark. He earned bachelor's and master's degrees in biomedical engineering at Shahid Beheshti Medical University and Polytechnic in Tehran, Iran, in 2002 and 2004, respectively. He is currently an associate professor in sports science and ergonomics at the Department of Health Science and Technology at Aalborg University, Denmark. He is the director of the Laboratory for Ergonomics and Work-Related Disorders. His specific research field is focused on methods of quantification of work exposure, risk factors for the development of musculoskeletal disorders and interactions between fatigue and motor control in various functional tasks, including computer work. The author has over 65 peer-reviewed journal articles, mostly related to the application of novel data analysis methods in the field of ergonomics and sport sciences. Dr. Samani serves as a reviewer in a number of journals within his fields of expertise and acts as an associate editor for the journal *Medical & Biological Engineering & Computing*. The author's teaching activity is focused on sport science and sport technology, as well as biomedical engineering and medical students.

1 Introduction

This book is not going to be a comprehensive reference for a signal processing course; however, it does present basic concepts discussed in almost all signal processing books. I put my emphasis on an intuitive understanding of what these concepts are about and show some examples of how they are being used in the scientific literature. I try to refrain from presenting mathematical formulations as much as possible; however, to get a good grasp of the concepts, some understanding of the math behind the topic is crucial; therefore, I have tried to keep it to a minimum level. I present some MATLAB[1] sample codes to make a case for the application of the concepts. MATLAB is a licensed software package, which may not be available for all readers of this book; in case readers do not have access to a license for MATLAB, I suggest using Octave,[2] whose syntax is largely compatible with MATLAB. However, I test all my sample codes in MATLAB and cannot guarantee that they are all translatable to Octave without any modification. I briefly introduce MATLAB in Appendix A.1, but I also encourage the reader to take advantage of the huge number of available online resources (mainly from MathWorks) providing instruction in MATLAB programming.

1.1 WHY DO WE NEED TO INTRODUCE THE THEORIES OF SIGNAL PROCESSING TO NON-ENGINEERS?

Like any field of science, signal processing and its rapid development are intertwined with the developments of other fields of technology and science. Modern signal processing technology emerged during the World Wars I and II, when scientists and engineers were dealing with radar and sonar signals and wanted to extract the signal from the background noise (Stillwell, 2013). The development of computers and the introduction of the digital world to various fields of technology, as well as the invention of new sensors enabling measurements of a wide range of physical quantities, allows signal processing to appear in various disciplines. It is obviously exciting to acquire objective measurements of a phenomenon under investigation, but this comes at a price, in that scientists have to deal with some sort of a signal and signal processing methods in their work. Thus, today signal processing appears in many scientific articles and in the methodological sections of the articles in one way or another. For example, I simply searched for two keywords, "EMG" which stands for electromyography and "sport" in the web of science.[3] The number of publications with these two keywords has increased markedly since 1994 (Figure 1.1).

Although sport science is a very broad field and many scientists in this field may have an engineering background, it is generally not known to be a branch of engineering science, and many scientists in this field do not necessarily have any engineering background. The EMG is a typical biological signal, and the aforementioned articles must inevitably address how EMG signals have been sampled, filtered and

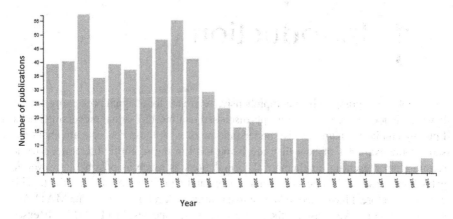

FIGURE 1.1　The number of publications with the keywords EMG and sport over the years

analyzed. All of these points have close ties to the theory of signal processing. If I repeat the same search with physiotherapy as the topic, a marked increase in publications from 1992 can be seen here too, but maybe not to the same extent as I could see for sport science. Obviously, this simple search does not qualify a scientific survey; nevertheless, it may be an indication of the widespread use of signal processing in various fields that are not necessarily known as branches of engineering science. A lack of basic knowledge in this field may cause fallacious interpretation of the results obtained in a scientific investigation. Therefore, gaining a basic knowledge of signal processing for non-engineers turns out to be very important.

1.2　WHAT IS A SIGNAL?

If one looks up the word in the Merriam-Webster dictionary,[4] there will be one definition that is very close to what is implied by a signal in the theory of signal processing, "a detectable physical quantity or impulse (such as a voltage, current, or magnetic field strength) by which messages or information can be transmitted". There is one very important keyword in this definition, and that is "information". However, a signal does not necessarily need to be a physical quantity; for example, a financial time series may contain the rating of a stock market over a certain period of time. Having said that, most often we deal with some sort of physical quantity in signal processing.

As mentioned earlier, "information" is the key word, as it provides the content of communication between humans or between humans and machines. Here, the term "machine" is being used in a broad context in which even the human body is a machine. For example, when we record biological signals (e.g., EMG) and process them, we essentially extract desired information from the complex machine of the human body.

1.3　WHAT IS NOISE?

When capturing signals in practical applications, what we acquire is not purely the signal (what we are interested in). There will be an undesired part that contaminates our signal and therefore is undesirable. This undesirable part is called "noise". Knowing the relationship between what we acquire, signal and noise is crucial to effectively remove the noise and keep the signal. For example, if one is interested in studying the level of muscle activity during a specific physical activity, the interference from the power line to the measured signal is not of interest, and therefore that interference is a part of the noise.

In many applications, a simple additive relationship is not quite far from reality, and very often this model is assumed to describe the relationship between what we measured (m), signal (s) and noise (n).

Thus, we may have:

$$m = s + n$$

This is called a model of an "additive noise", meaning that the noise is simply added to the signal, and the measurement is simply the summation of the noise and the signal. However, in a general case, the relationship between signal and noise can be more complex, and the measurement can be an unknown function of the signal and noise. In an experimental design, special precautions are taken such that the noise is as minimal as possible, and an additive model of noise could be often assumed.

In certain cases, even though the noise model is not essentially additive, using a little mathematical trickery, an equivalent additive noise model can be found. For example, if $m = s.n$ (m equals s times n), the noise is productive, but if we simply take the logarithm of both sides of the equation, this case can be a transformed into an additive noise model as $\log m = \log s + \log n$.

The quality of the measurement refers to a question about how big the noise term in the equation is with respect to the signal term. In technical documents and papers, one may come across a term called signal-to-noise ratio (SNR), which simply carries this piece of information about the magnitude of a signal with respect to the noise. In physics, this is expressed in terms of the ratio between the power of the signal and noise, and because the range of signal and noise power can be quite wide, a logarithmic scale is used. The power of the signal and noise implies how much energy in a unit of time is being transferred by the signal and noise.

Thus:

$$SNR = 10.log_{10}\left(\frac{P_{signal}}{P_{noise}}\right)$$

As the signal and noise are often measured in voltage or current, it may seem more convenient to express the SNR in terms of the magnitude of voltage or current. If one remembers the basic physics of electrical circuits—for example, for a resistor—the power is proportional to the square of voltage across the resistor or the current passing through it. Thus, if the SNR is calculated in terms of voltage or current amplitude, one can write:

$$SNR = 10.log_{10}\left(\frac{A_{signal}^2}{A_{noise}^2}\right) = 20.log_{10}\left(\frac{A_{signal}}{A_{noise}}\right)$$

This relative index is expressed in decibels (db) and because, in a general case, the current or the voltage varies across time, the amplitude is calculated in terms of the root mean square (RMS). RMS of a signal in a limited time window is calculated by taking the average of the samples of the signal to the power of 2 and then taking the square root of the average.

$$RMS = \sqrt{\frac{1}{N} \sum_{i=1}^{N} x_i^2}$$

Where N is the number of signal samples in the limited time window and x_i is the i-th sample of the signal in that time window. The notation $\sum_{i=1}^{N}$ simply means a summation when i changes from 1 to N. For example, a 20-db SNR implies that the RMS of the signal is 10 times larger than that of the noise; similarly, a 0-db SNR means that signal and noise have equal RMS, and negative values of SNR mean that the noise is even greater than the signal. SNR in biological signals such as EMG can be estimated based on a part of the signal spectrum (this concept will be introduced in chapter 6), which is expected not to contain the signal power related to the signal (Sinderby, Lindstrom, and Grassino, 1995).

Now, I established a very brief definition of what signal and noise are in general terms. Throughout the rest of this book, first, I introduce the main building blocks for recording signals, and then I continue with an introduction to the time and frequency domains of the signals, and a brief introduction to the sampling theorem and power spectrum of signals will follow. After touching on the most basic topics related to the signals, discuss systems that include, for example, filters and any blocks acting on input signals and that produce output signals. When I provide examples throughout this book, I mostly refer to examples in the field of biological signals, kinesiology, sport science, biomechanics and ergonomics. However, the main part of what I write here encompasses general concepts of signal processing, which are to a large extent applicable to other fields.

NOTES

1. These MATLAB files are available at https://www.crcpress.com/9780367207557.
2. www.gnu.org/software/octave/
3. http://apps.webofknowledge.com/WOS_AdvancedSearch_input.do?SID=C2Ykq83coa AC4oJ92Ru&product=WOS&search_mode=AdvancedSearch, (TS=(EMG AND sport)) AND LANGUAGE: (English) AND DOCUMENT TYPES: (Article)
4. www.merriam-webster.com/

REFERENCES

K. Bromley and H. J. Whitehouse, "Signal processing technology overview," *Real-Time Signal Process. IV.*, vol. 298, pp. 102–107, 1982.

C. Sinderby, L. Lindstrom, and A. E. Grassino, "Automatic assessment of electromyogram quality," *J. Appl. Physiol.*, vol. 79, no. 5, pp. 1803–1815, 1995.

2 The Measurement Pipeline

Working with signals involves some sort of measurement. Sometimes this measurement does not require any complicated hardware—for example, if one wants to record the worth of the stock market over a certain period, access to the registered values of the stock prices may be sufficient to start the work. However, recording biological signals (bio-potentials), such as electromyography (EMG), first requires some hardware in order to capture the signal. Here, I am going to briefly describe the basic blocks of the measurement systems that we often use in the lab to record bio-potentials. If the reader is already familiar with these basic concepts, this chapter can be skipped over.

2.1 SENSORS

Sensors are essentially some sort of a convertor of various energy forms to electrical energy (Webster, 1998) (page 6 therein). Why are we even talking about energy forms? Very often, the pieces of information lie in physical quantities, which are produced by transforming energy from one form to another. For example, changing the velocity of an object requires that some source of energy be transformed into mechanical energy. To measure the velocity, we generally prefer to convert the mechanical energy to electrical energy because handling electrical energy is easier for us. Given the widespread use of computers, it is easier for us to store the measured values on the computers and process them (Areny and Webster, 2001). Therefore, the development of sensors would depend on understanding the physics relating to the variations of a "measurand" (what is to be measured) to an electrical quantity (charge, current or voltage). For example, if the displacement of an object is what one wants to measure, the displacement may be linked to the resistance, capacitance or inductance of an element in an electrical circuit. Then, the circuit can be designed such that the voltage across that element or the current which passes through it is proportional to the amount of displacement. Thus, measuring the voltage or current enables measuring the displacement. The sensor world is a very huge and exciting world. Depending on their precision, linearity, range of work and other specifications, sensors can be quite cheap or very expensive and strategic.

2.2 AMPLIFICATION

Very often, the output of sensors is a weak electrical voltage (it could be an electrical current too, but often we have to deal with voltage signals), so in order for us to be able to process and store it on a computer, higher voltage is required. Therefore, ideally speaking, an amplification block is a block to multiply the signal by a constant factor. To be more precise, we may also have amplification with an automatic gain control in which the gain can be changed depending on the voltage level of an input signal, but in many experimental devices to record biological signals, we use an amplifier with a constant "gain". Figure 2.1 shows a typical symbol that is drawn on the schematic of electrical circuits to show an amplifier with a constant gain of "K". Thus, if the input to the amplifier is x, then the output would simply be K.x (K times x).

The amplifier used to record biological signals is a specific class of amplifier, which is known as a "differential amplifier" (Figure 2.2). This type of amplifier has two input leads, and the output is the amplification of the difference between the two input leads. In this case, the output is simply $K(x^+ - x^-) = K \cdot V_d$ where V_d is the differential input.

The input voltage on each of the input leads (i.e., x^+, x^-) can be reformulated as:

$$x^+ = \underbrace{\left(\frac{x^+ - x^-}{2}\right)}_{V_d^+} + \underbrace{\left(\frac{x^+ + x^-}{2}\right)}_{V_c}$$

$$x^- = \underbrace{\left(\frac{x^- - x^+}{2}\right)}_{V_d^-} + \underbrace{\left(\frac{x^+ + x^-}{2}\right)}_{V_c}$$

It is very straightforward to verify that $(x^+ - x^-) = (V_d^+ - V_d^-) = V_d$. This is the differential input to the amplifier, but as one can see, there is another term in the equation, that is V_c being a common term for x^+ and x^-. This term is called the "common

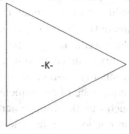

FIGURE 2.1 A typical symbol of an amplifier with a gain of K on a schematic of electrical circuits

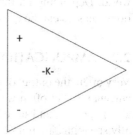

FIGURE 2.2 A typical symbol of a differential amplifier with a gain of K on a schematic of electrical circuits

mode" input. In an ideal scenario, this term is irrelevant because it does not affect the difference between x^+ and x^-. However, in a practical case, this term is quite relevant, as I explain next.

2.2.1 COMMON MODE REJECTION RATIO

In a practical differential amplification, what one gets in the output is not only the amplification of the difference between the two input leads but also some fraction of the common mode that contributes to the output. Intuitively, one expects that the common mode would be amplified to a much lesser extent compared with the differential input so that the response of a practical amplifier would be close to its ideal version. This is actually the case, and the differential gain is much greater than the common mode gain—but how much greater? This is what is being represented by the common mode rejection ratio (CMRR).

Imagine that the differential gain is K_d (the gain applied to the differential term) and the common mode gain is K_c (the gain applied to the common mode), then the output will be $K_d \cdot V_d + K_c \cdot V_c$. In a similar fashion, as one formulates for the signal-to-noise ratio (SNR), the magnitude of K_d with respect to K_c is expressed in a logarithmic scale. Thus,

$$CMRR = 20 . log_{10}\left(\frac{K_d}{K_c}\right)$$

The established guideline of data recording urges scientists to notice these technical characteristics of their experimental setup and report them in their publications (Merletti and Torino, 1999). Often, the CMRR should be greater than 80 db.

A practical example of the relevance of the CMRR may be when one records a bipolar surface EMG from the upper trunk (e.g., trapezius activity). Two EMG electrodes are placed on the muscle, and they are supposed to capture the electrical activity underlying the electrodes and then differentially amplify them. However, because the site of recording is close to the heart, the electrical activity of the heart also contributes to what each electrode captures. Part of this contribution is common on both EMG electrodes, which can be seen as the common mode. One expects that this common mode will be removed when the EMG is amplified differentially; however, it is quite common for the heart's electrical activity to be seen on the recorded EMG, and this source of contamination should be specifically removed (Marker and Maluf, 2014) (Figure 2.3). Note that what can be seen in Figure 2.3 is not solely due to the amplification of the common mode, as the interference of the heart rate is not exactly the same across the EMG electrodes, and a big fraction of the peaks seen in Figure 2.3 is simply due to the amplification of the differential component.

Another example may be the interference of the power line on the measurement of a biological signal (say EMG). When electrical circuits are fed by the power line, an alternating current with 50 Hz (in North America 60 Hz) is flowing in cables and wires, and this may cause electromagnetic interference with a measurement

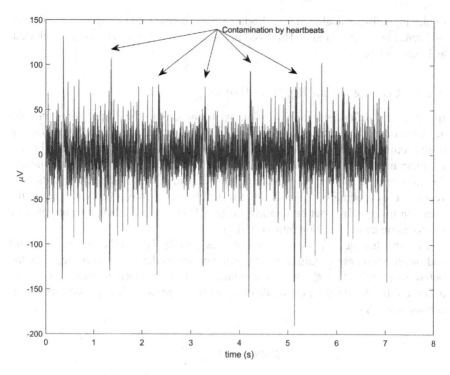

FIGURE 2.3   A typical recording of bipolar EMG on the lower trapezius at the resting state; interfering heart activity on the EMG signal is quite visible

setup. Depending on the quality of isolation and the grounding of the measurement devices, this interference can be considerable, and sometimes a notch filter (defined in Chapter 8) is applied to the signal to reduce the effect of this interfering noise. The interference effect on each of the electrodes is supposedly the same; thus, ideally, the differential amplification should cancel this term, but in some experimental conditions, the interference may still be seen.

2.2.2   How Much Can the Signal Be Amplified?

From the previous discussion, one may suppose that an amplifier simply multiplies a constant to an input signal and provides an amplified version of the input in the output. Is there any limit? Can the output voltage of an amplifier get any amplitude? Of course, there is a limit. One of the main determinants of this limit is the supply voltage of the amplifier. For the amplifier to work, it should be supplied to allow for a change in the voltage level in the output. The output voltage of the amplifier cannot be greater than its supply voltage. If the amplifier is battery powered, one may think that the supply voltage of the amplifier equals the maximum voltage that one can get from the battery at maximum. If the gain or the input voltage level is too high such that the simple multiplication of the gain and the input voltage exceeds the supply voltage, output voltage cannot exceed a certain limit that is lower than the supply voltage.

2.2.3 INPUT IMPEDANCE

I assume most readers of this book are familiar with the term of resistance in electrical circuits from high school physics. Resistance refers to opposition against the electrical current in an electrical circuit. Again from high school physics, one could remember the relationship between resistance, current and voltage, known as "Ohm's law", is voltage across a resistor (V) equals the multiplication of the electrical current passing through it (I) and its resistance ($V = R.I$).

The term impedance is a generalized concept of resistance and can also refer to electrical elements other than resistors, such as capacitors and inductors. Unlike resistors, which consume active power, capacitors and inductors only store energy; nevertheless, they can affect the relationship between voltage and current. For the simplicity of the text in this chapter, whenever I refer to impedance, the reader may assume that resistance and impedance are referring to the same concept, but to be precise these two terms are not the same thing.

Now imagine that one has an electrical circuit shown in Figure 2.4 and wants to measure the voltage across a resistance (R2). The voltmeter is placed parallel to R2. In an ideal case, the measurement device should have no effect on the main circuit, as if the measurement device is not connected. In practice, the measurement device has an effect on the main circuit, but this effect should be minimized. What should be done to reduce the effect of the measurement device on the main circuit? If the resistance of the voltmeter against the electrical current is much higher than that of R2, the current that passes through the voltmeter will be negligible in comparison with the current flowing through R2, and one could assume with reasonable certainty that the measured voltage across R2 is very close to the voltage without the presence of the measurement system. The resistance of the measurement system against the electrical current is known as the input impedance.

Because biological signals are voltage signals in many cases, the story of their measurement resembles the previously mentioned scenario. One does not want the amplifier system to interfere with the recordings; therefore, the input impedance should be high. Very often, the input impedance is greater than 1 megaohm.

There are other important properties of an amplifier, like linearity and the bandwidth, which will be addressed when system properties are discussed in Chapter 7.

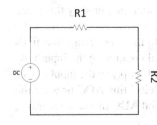

FIGURE 2.4 Schematic of an electrical circuit. The setup to measure voltage across a resistor (R2) must be parallel to R2

2.3   ANALOG-TO-DIGITAL CONVERSION

Let us imagine that one has amplified a signal properly and now needs to store it on a computer or wants to do some real-time analysis of the signal on the computer. How does one access the signal in a computer? These days many people are familiar with analog-to-digital convertors (ADC), but let us assume that we did not know about them. How would one intuitively store signals on a computer? Once I asked this question of my students in a lecture, and one of them responded, "I touch the signal; if it hurts, it is high. Otherwise, it is low". I replied, "Well, I do not recommend doing it like that, but conceptually you are essentially not horribly wrong". Next, I outline the function of an ADC, but what follows is not an exact description of what is actually implemented at the hardware level of an ADC.

The first step would be to have a bunch of voltage thresholds and compare the input voltage level with those thresholds. Say one has 10 voltage thresholds sorted ascendingly. Now imagine if one input voltage is higher than the third threshold and lower than the fourth one. One could then infer that the input voltage is between the third and fourth voltage thresholds. Now the question is this: How much is the difference between the third and fourth thresholds? If the difference is very low, we have actually measured our input voltage with an acceptable level of uncertainty. A term of uncertainty is an inherent part of any measurement, and the difference between the two threshold levels in an ADC could also be counted towards the uncertainty of measurement. Suppose that one is measuring an input voltage that is about 1 mV and the difference between two successive thresholds is about 2 µV—for example, the voltage of the third threshold is 999 µV and the fourth threshold is 1001 µV. Having found that the input voltage is between the third and fourth thresholds means that the input voltage is something between 999 and 1001 µV. Finding this information is very important to us because we may not care about 2 µV of uncertainty when working with a voltage of 1 mV. In other words, the resolution of the measurement is sufficiently fine.

Now imagine that one knows that the input voltage is within a specified range. For example, one may know that the amplified signal will never get an amplitude outside the range of ±5 V. This piece of information is crucial to have when working with a specific type of signal. Often, this information can be found in the scientific literature investigating that specific type of signal. For example, surface EMG can vary from ±5 mV (Konrad, 2005) (including extreme cases among athletes), and if the applied amplification gain is 1000, one knows that the input range will be ±5 V. If one works with another type of signal, this information should be retrieved from the literature. If one starts working with a type of signal with no prior information, one can start from a very conservative guess of the range and then modify this piece of information with experience.

In the first step, the ADC applies a set of thresholds to the input range and finds the two successive thresholds, which determine an interval containing the input voltage. Now the question is how many thresholds does an ADC apply to the input range? The ADC datasheet contains information on how many "bits" this ADC uses for the conversion. For example, one may use a 12-bit ADC or 16-bit ADC in a data recording scenario. This number determines how many intervals are between the minimum and maximum of the input range. If one has an ADC with n-bits to do the conversion, there will be $2^n - 1$ number of intervals between the lowest and highest amplitudes in the input range. For example, in case of the EMG outlined earlier, if one uses a 16-bit ADC, there will be $2^{16} - 65535$ intervals, and the measurement resolution for the amplified signal will be 152 μV, which results from the division of the input range (10 V) by 65535. Now we can code the input voltage based on the intervals; for example, the 1000th interval could correspond to $-5V + (1000 \times 152$ μV$) = -4.848V$. Instead of storing the voltage values directly, one can now simply store the coded intervals.

Now we are faced with the second challenge: How do we store these numbers on a computer? I assume many readers of this book have heard that computers can only store two digits, namely, 0 and 1. Then how can a number with different digits from 0 and 1 be stored in a computer? For example, if the ADC conversion in the earlier example ended up with a coded interval of 598, how would these numbers be stored on a computer? The answer to this question is quite straightforward. The reason for having numbers with digits other than 0 and 1 is that the regular (Arabic) numbers are presented in the base of 10. If one presents the numbers in the base of 2, the numbers can be represented by only two digits of 0 and 1 (also known as binary digits).

A regular number is a sequence of digits, and each digit also has a place-value in the sequence, and the place-value is determined by the order of the digit in the sequence. The digits on the left side of the digit sequence have a higher place-value, which is a kind of weighting. Imagine of a sequence of digits corresponding to the number 598, which can be written as the summation of all the digits multiplied to 10 powered to its place-value.

$$\begin{matrix} place-value & 2 & 1 & 0 \\ a\ number\ in\ base\ of\ ten & 5 & 9 & 8 \end{matrix} = 5\times10^2 + 9\times10^1 + 8\times10^0 = 598$$

The same story can be told about the numbers in the base of 2. For example, if a number in the base of 2 is given, it can be converted to a number in the base of 10 following the same approach outlined earlier.

$$\begin{matrix} place-value & 2 & 1 & 0 \\ a\ number\ in\ base\ of\ two & 1 & 0 & 1 \end{matrix} = 1\times 2^2 + 0\times 2^1 + 1\times 2^0 = 5$$

Thus, if the numbers are presented in the base of 2, it will be straightforward to store them on a computer, and the computer can read and analyze the numbers.

Thus, intuitively speaking, an ADC performs these two steps. First, it splits the input range in equidistance intervals and finds the interval containing the input. Then, it finds the corresponding binary code for that interval and can store the input value and use it in computational work. The algorithm that is often used in ADC to perform this task is called successive approximation. Figure 2.5 shows the building blocks of a typical measurement setup in the lab.

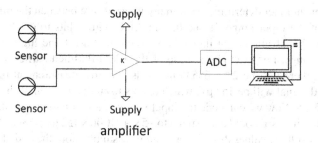

FIGURE 2.5   Building blocks of the measurement pipeline in a typical experimental setup

In this chapter, I briefly explained the main building blocks in a measurement pipeline. When running an experiment in the lab or in the field, one will need to work with some type of a sensor, an amplifier and an ADC to capture and store the dataset. In case the experiment involves online processing of the dataset, analysis should also be conducted on the dataset, and a specific output will be generated.

REFERENCES

R. P. Areny and J. G. Webster, *Sensors and Signal Conditioning.* USA: John Wiley & Sons, Inc, 2001.

P. Konrad, "Noraxon: The ABC of EMG," *Signal Processing.* USA: Noraxon U.S.A, Inc, 2005.

R. Merletti and P. Di Torino, "Standards for reporting EMG data," *J. Electromyogr. Kinesiol.,* vol. 9, no. 1, pp. 3–4, 1999.

R. J. Marker and K. S. Maluf, "Effects of electrocardiography contamination and comparison of ECG removal methods on upper trapezius electromyography recordings," *J. Electromyogr. Kinesiol.,* vol. 24, no. 6, pp. 902–909, 2014.

J. G. Webster, *Medical Instrumentation Application and Design.* USA: John Wiley & Sons, Inc, 1998.

3 Time and Frequency Representation of Continuous Time Signals

Before getting started with this chapter, I have to clarify a few terms used in its title. The first question may refer to the concept of a "continuous time signal". I have already made a brief definition of a signal and what it represents in this book. But what is a continuous time signal? By definition, this means that for any given time instance, no matter how fine the temporal resolution is, a continuous time signal has a corresponding amplitude. Mathematically speaking, a continuous time signal is a function of time and has an amplitude for any given time point. At the first glance, this may seem like a very trivial property, but when we discuss discrete time signals, it can be appreciated why one needs to distinguish continuous time signals.

Any physical signal around us is a continuous time signal. Whether it be communication signals, electromagnetic waves, a biological signal or any biomechanical signal, they are all continuous time signals. The second question is about the "time representation". When one looks at the variation of a signal across time, it is the time representation of the signal that is seen. Mathematically speaking, the signal is then a function defined in the domain of time. For example, Figure 2.3 shows a short segment of an EMG signal recorded from the lower trapezius muscle. As can be seen, the abscissa (X-axis) in Figure 2.3 represents the time, and the ordinate (Y-axis) represents the corresponding amplitude of the signal for each time point.

I still have to clarify what I mean by the frequency representation, and this is not as straightforward as the time representation because people have a much clearer mental image of the time representation of a signal compared with its frequency representation. Let us take a second and think about our understanding of the concept of frequency. What do we understand about it? When I ask this question of my students, they often respond with terms such as "the rate of repetitions" or oscillations, which are not quite that far from what the concept of frequency actually implies. Many people have at least heard of the frequency unit Hertz (Hz), and in their minds, Hz is associated with frequency. If I ask students about their impression of a specific frequency, let say, 50 Hz, they often describe a mental picture of a repeating pattern every 20 milliseconds (ms) which is the reciprocal of 50 Hz (50 Hz = 1/20 ms). Thus, they essentially think of periodic signals when they think of the concept of frequency.

3.1 PERIODIC SIGNALS

What is a "periodic signal"? Intuitively speaking, if the signal amplitude repeats itself on certain intervals across time, that signal is called periodic. That definition may be understandable by some people, but it is quite vague. Imagine a signal that has a constant amplitude (e.g., a DC voltage). The signal amplitude repeats itself all the time, but this is not what one would call a periodic signal. The mathematical definition is very clear, and yet it is still quite straightforward to understand without being too complicated.

Imagine a signal $x(t)$ represented in time, if for any arbitrary time point t, there exists a constant greater than zero ($T > 0$) such that $x(t) = x(t + T)$, then $x(t)$ is called a periodic signal. T is called the period of the periodic signal, but is it a unique number? No, because if T is a period satisfying the condition outlined earlier, two times T ($2.T$) and three times T ($3.T$) and essentially all the integer multiples of T would also satisfy the condition. However, the smallest value of T that satisfies the condition is unique, and it is called the "main period" of a periodic signal. Hereafter, whenever I refer to the period of a periodic signal, I refer to the main period unless otherwise stated. Figure 3.1 shows some typical examples (triangular, square and sine waves) of what one can call periodic signals.

As can be seen in Figure 3.1, triangular, square and sine waves all can have the same period of T, but then does it mean that they all have the same frequency representation? Intuitively speaking, they must have different frequency representations; otherwise, one cannot distinguish them in the frequency domain. Imagine that in Figure 3.1 the abscissa (X-axis) represents the frequency axis: How should we represent the periodic signals depicted in Figure 3.1?

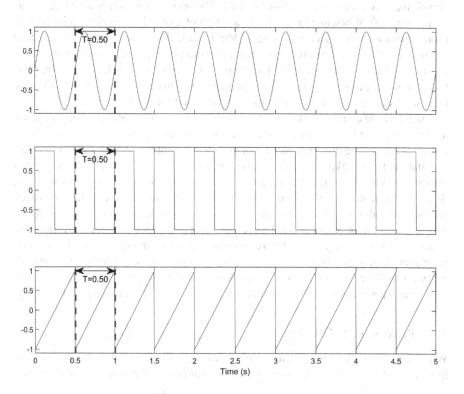

FIGURE 3.1 A few examples of periodic signals with the same period of $T = 0.5$ second

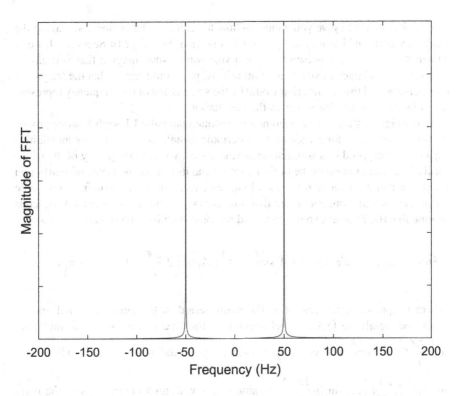

FIGURE 3.2   Frequency representation of a sine wave signal

In case of a sine wave, it is easier to depict that because we can show it as a very sharp impulse in the frequency domain. This means that if we have the frequency axis on the X-axis, we should plot a sharp pulse at the reciprocal of its main period to represent a sine wave in the frequency domain. It must have another pulse on the "negative" frequency (see Section 3.3) axis too, but let us not bother with that part for the moment. Basically, one has to be familiar with the mathematical foundation of the frequency representation to realize why a sine wave can be represented by a single pulse in the frequency domain; however, for now, we will keep the discussion simple. Figure 3.2 shows the frequency representation of a sine wave with 50 Hz. Thus, a sine wave is called "monochromatic" in the frequency domain because it corresponds to a single positive frequency point on the frequency axis.

Then how about the triangular and square waves? How do we then represent them in the frequency domain? Before answering this question, we need to familiarize ourselves with a mathematical concept called Fourier series. I promised not to get involved in the math, and I will do my best to keep this promise, so in Section 3.2, I will try to outline its concept as straightforwardly as possible.

3.2   FOURIER SERIES

What if I tell you that if you give me a periodic signal, I can then expand it as a summation of a bunch of sine waves with different periods and weighting factors—don't you think that would give you a clue on how to represent the periodic signals in the frequency domain? I would say yes to this question, but why? In Section 3.1, I discussed the frequency representation of a sine wave. Now imagine that you have a summation of a bunch of sine waves; intuitively, you would expect that the frequency representation of this summation would be the summation of the frequency representation for each of the sine waves in the summation.

In the eighteenth century, a French mathematician called Joseph Fourier proved (Stillwell, 2013) that for a wide class of periodic signals, one could write an additive expansion composed of a summation of sine waves, with the frequency of the reciprocal of the main period of the periodic signal and its integer multiples with different weighting factors. I know this was a long sentence, but let me explain. I will write the proper formulation, but before that, just to have a little simple formulation, let us assume that the Fourier expansion would be something like the following:

$$An\ arbitray\ periodic\ signal = A_1 \sin\left(\frac{2\pi}{T}t\right) + A_2 \sin\left(2.\frac{2\pi}{T}t\right) + \ldots + A_n \sin\left(n\frac{2\pi}{T}t\right)$$

Where t represents the time, T is the main period of the periodic signal and all A_i are the weighting factors (coefficients) for the sine waves. In this formulation, $\sin\left(\frac{2\pi}{T}t\right)$ is a sine wave whose main period equals that of the given periodic signal, and $\sin\left(2.\frac{2\pi}{T}t\right)\ldots\sin\left(n\frac{2\pi}{T}t\right)$ are sine waves with multiple integers of the main period. Given this summation, one can simply imagine the frequency representation of this arbitrary period signal would look like what is shown in Figure 3.3.

How does one find the weighting factors? And how many should there be? To answer these questions, one has to seek out the full formulation. Actually, the full formulation is slightly more complex, so here I cannot fully keep my word to not get involved with math, but I will try to keep it as minimal as possible. If you do not like to see any mathematical notation, just skip the next few lines. For a given arbitrary periodic signal $x_P(t)$ with a main period of T, the $x_P(t)$ can be expanded as:

$$x_P^M(t) = a_0 + \sum_{i=1}^{M} a_i \cos\left(\frac{2\pi}{T}i.t\right) + \sum_{i=1}^{M} b_i \sin\left(\frac{2\pi}{T}i.t\right)$$

$$Where\ a_0 = \frac{1}{T}\int_0^T x_p(t)dt,\ a_i = \frac{2}{T}\int_0^T x_p(t)\cos\left(\frac{2\pi}{T}i.t\right)dt\ and\ b_i = \frac{2}{T}\int_0^T x_p(t)\sin\left(\frac{2\pi}{T}i.t\right)dt$$

In this notation $\sum_{i=1}^{M}$ represents a summation across i from 1 to M and $\int_0^T$ is a definite integral over the main period of the periodic signal $x_P(t)$. If you are not familiar with the concept of integrals, imagine it as a summation of incremental parts under a curve to obtain the area under the curve.

If one pays attention to this formulation they will note that I intentionally used $x_P^M(t)$ and not $x_P(t)$ itself. So what is $x_P^M(t)$? In order for the summation provided

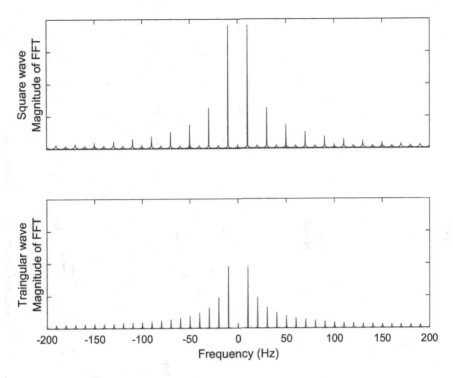

FIGURE 3.3 The frequency representation of an arbitrary period signal as outlined in the equation provided

earlier to tend to $x_p(t)$, the M should go to infinity, but as M increases, the weighting factors a_i and b_i get smaller, and their contribution to the summation will be less and less. Therefore, in many practical applications, it will be sufficient to truncate the M to some fixed number and discard the negligible remaining terms in the summation. Thus, initially as M increases, the summation becomes closer and closer to the actual periodic signal. Figure 3.4 shows the effect of increasing M when a square wave signal is reconstructed by a limited number of M. As is clear from Figure 3.4, the reconstructed signal from the summation of the Fourier series converges to the original periodic signal as M increases. However, if one pays close attention, it is not too difficult to see that at the points of "discontinuity" in the original signal, no matter how one increases the M, the reconstructed signal still exhibits a large overshoot with respect to the original signal. This is called the Gibbs phenomenon, and it is because the Fourier series is the summation of a limited number of continuous functions, and therefore the resulting sum has to be continuous, whereas the original signal has a discontinuity. As M increases, the location of the overshoot approaches the exact location of discontinuity but the overshoot never disappears.

In this notation, the sine and cosine waves with the frequency of the multiple integers of the main frequency are called harmonics. For an even i (2, 4, 6, . . .), they are called even harmonics, and for odd i (1,3,5, . . .), they are called odd harmonics. Figure 3.5 shows how one can computationally derive a_i and b_i weighting factors in

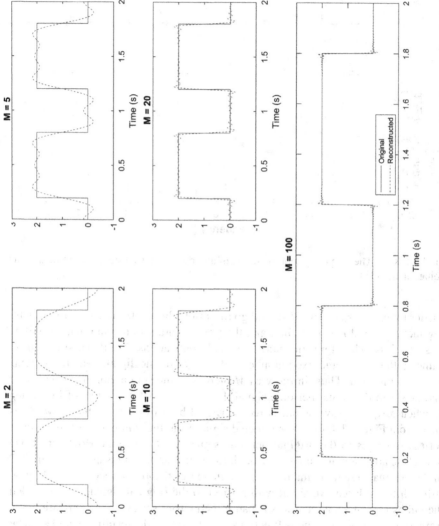

FIGURE 3.4 The Fourier series summation for $M = 2$, 5, 10, 20 of a square wave signal over two periods

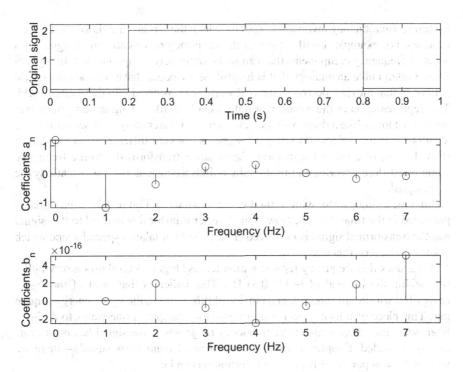

FIGURE 3.5 The frequency representation of square and triangular waves derived from Fourier series weightings

the notation in MATLAB[1] for a square wave. Clearly, the frequency representation of these signals is not limited to one single frequency point, but rather it consists of various frequency components. Thus, the signal is not monochrome in the frequency domain.

3.3 FOURIER TRANSFORM

So far I only addressed periodic signals, but how about non-periodic signals? This is a valid question because in practical applications, we frequently deal with non-periodic signals. Actually, the concept of a Fourier series can be generalized to non-periodic signals to derive a frequency representation for this type of signal. A Fourier transform is a mathematical transform that allows one to derive a frequency representation for these signals. The signal should satisfy certain criteria known as Dirichlet conditions before the Fourier transform can be applied. Luckily, many signals that we work with satisfy these conditions. The concept of mathematical transformation may be not familiar for many readers of this book. In signal processing, it is very common to apply various forms of transformation to signals, because the transformed domain of the signals has some features that facilitate the processing and provide the required information. Sometimes, the transformed domain creates an intuitive impression of what the domain is about—for example, in the case of the frequency domain, people have a sense about what the frequency may imply. However, the transformation may involve concepts for which there is no straightforward understanding. For example, I will show that the frequency representation of signals has negative frequency components that cannot be intuitively comprehended. In my lectures, I often make an analogy that is helpful. Many people have watched the movie *Avatar*. In that movie a person in the real world is transferred into the Avatar world. This representation in the Avatar world does not look like a human being anymore, but instead looks like a dragon-shaped creature yet it essentially represents the same identity. The story of transforming the signal is not very different from the Avatar story. In case of a Fourier transform, the signal is transformed into the frequency domain (a different world). In this domain, it finds a new look but it is essentially the same signal.

Imagine a single pulse with limited width as shown in Figure 3.6a—the signal is presented in the time domain. Now if the Fourier transform is applied to this signal and the transformed signal is presented (Figure 3.6b), it takes a special shape which is known as a sinc function.

What does this frequency representation tell us? Figure 3.6b shows a very bulky part within the interval of (−0.5, 0.5) Hz. This indicates that most of the signal energy lies within this interval, and outside of this interval, the signal energy is quite low. This piece of information is important—for example, if one wants to design a filter, some information about the frequency range where the signal has most of its energy is needed. Chapter 4 discusses sampling of continuous signals—there we will see how important this piece of information can be.

Let us consider another example. Figure 3.7a shows a damping exponential signal. It is also called free induction decay and can be recorded using nuclear magnetic resonance spectroscopy. Figure 3.7b shows the magnitude of its corresponding Fourier transform. As opposed to the time representation of the signal, which is extended over a period of time, its frequency representation is quite condensed in a narrow frequency range.

Here I intentionally used the term "magnitude" of a Fourier transform. What does the magnitude of Fourier transform mean? As I mentioned at the beginning of this section, a Fourier transform is a mathematical transformation, and the transformation results in a complex number in many cases. For those who are not familiar with complex numbers, there is a brief introduction in Appendix A.2. The numbers that we work with on a daily basis are called real numbers, but complex numbers have two different components (i.e., magnitude and phase) and are very often used when analyzing a phenomenon with two independent components. Luckily, in this case, we have a good understanding about what each component represents. The magnitude of a Fourier transform shows how the energy of a signal is distributed across the frequency axis, whereas its phase carries some information about a shift in time, meaning that if two signals are identical except that one is shifted with respect to the other signal in time, their difference in the frequency domain will be reflected in the phase of the Fourier transform.

Figure 3.8 clarifies this point. Two signals are shown in Figure 3.8a that are identical but are simply shifted in time. Figure 3.8b shows that the magnitude of the Fourier transforms of these two signals are identical, but Figure 3.8c shows that the phases of the Fourier transform are different, and a clear difference between the signals can be seen.

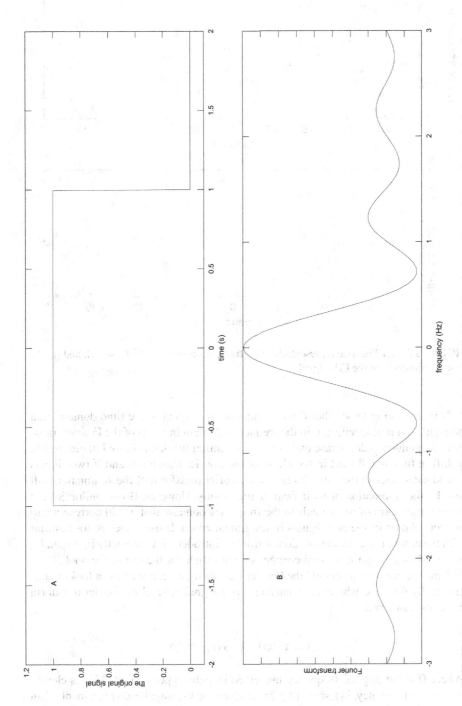

FIGURE 3.6 (a) The time representation of a single pulse in time, and (b) the Fourier transform of that pulse in the frequency domain

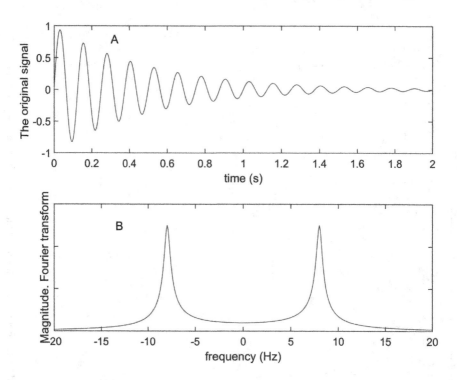

FIGURE 3.7 (a) The time representation of a free induction decay (FID) signal, and (b) the Fourier transform of the FID signal

It is interesting to see that if we manipulate the signal in the time domain, that manipulation is also reflected in the frequency domain in terms of the Fourier transform magnitude and/or phase and vice versa. Similar to what I showed in Figure 3.8, a shift in time is reflected in the phase of the Fourier transform, and if two signals add to each other in the time domain, the Fourier transform of the summation will also be the summation of their Fourier transforms. However, if one multiplies the Fourier transform of two signals in the frequency domain, that would correspond to the convolution of the two signals in the time domain. If the reader is not familiar with the concept of convolution, do not worry. I introduce it very briefly in Appendix A.3. Understanding the general concept should suffice for the rest of the book.

Now that we have discussed the Fourier transform, we can see how it looks mathematically for those who may be curious. For a signal $x(t)$ and its Fourier transform $X(\Omega)$, one can write:

$$x(t) \xrightarrow{\mathcal{F}} X(i\Omega) = \int_{-\infty}^{\infty} x(t) e^{-i\Omega t} dt$$

Where Ω is the angular frequency presented in radians per second. This is closely tied to the frequency, but scaled by 2π to resemble the angular displacement of an object rotating around a canonical point in 1 second. i is the imaginary unit of a

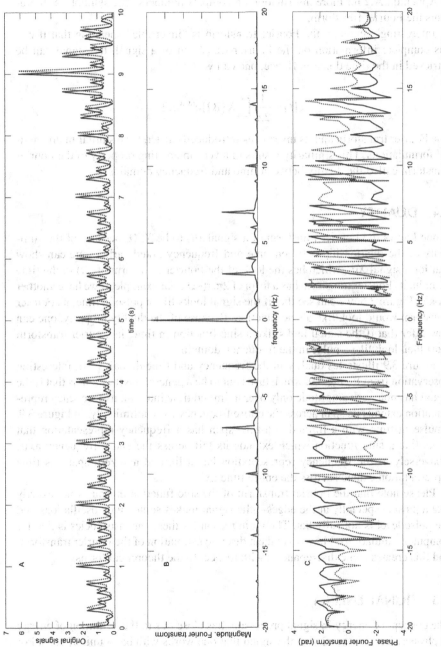

FIGURE 3.8   (a) The time representation of two signals with only a minor time shift difference, and (b) the magnitude of the Fourier transforms of two signals and (c) the phase of the Fourier transforms of the two signals

complex number $\left(i = \sqrt{-1}\right)$, as I mentioned earlier. The reader is advised to refer to Appendix A.2 for more information on complex numbers. The symbol $\xrightarrow{\mathcal{F}}$ represents the Fourier transform.

Interestingly enough, the Fourier transform is "invertible", meaning that if one has complete information on the Fourier transform of a signal, the signal can be retrieved in the time domain. Hence, one can write:

$$x(t) = \frac{1}{2\pi}\int_{-\infty}^{\infty} X(i\Omega)e^{i\Omega t}d\Omega$$

The Fourier transform and its inverse, as introduced earlier, have similar mathematical formulations. This similarity results in a very interesting property of the Fourier transform called the duality between time and frequency domains.

3.4 DUALITY

If one knows the Fourier transform of a signal $x(t)$ to be $X(i\Omega)$ because of the similarity of the relationship between time and frequency noted earlier, one can show that for a signal $X(t)$, which has the look of the Fourier transform of $x(t)$ in the time domain, its Fourier transform has a form of $2\pi.x(-\Omega)$. For example, if we have another look at Figure 3.6, we can see that if the signal looks like a pulse in time, its Fourier transform would look like a sinc function. Because of the duality property, one can now infer that if the signal looks like a sinc function in time, its Fourier transform must then look like a pulse in the frequency domain.

Figure 3.9 illustrates duality in the frequency and time domains. An interesting observation from Figure 3.9, which turns out to be a general observation, is that if the signal has non-zero amplitude only over a limited time interval, its frequency representation extends its span across the entire frequency axis unlimitedly. In Figure 3.9, a pulse in time with a limited non-zero span has a frequency representation that looks like a sinc function which extends its tail across the entire frequency axis. Conversely, if the frequency representation has a limited non-zero span, its time representation extends across the entire time axis.

Please note that the Fourier transform of the sinc function in time is not exactly like a pulse, especially at the edges—the signal makes some quick oscillations and the pulse level is not fully flat. The main reason for these inconsistencies is that for computer simulations, I have to make a discrete estimation of the Fourier transform, and this creates some discrepancies with respect to the theoretical forms.

3.5 SIGNAL ENERGY

The concept of energy in signal processing has close ties to the definition of energy in physics. Let us assume that the signal that one works with has a unit of volts. For example, it could be bio-potential signals such as an EMG. If one remembers high school physics, the energy can be obtained by the integration of power over time. For a voltage signal, the power would be the voltage squared divided by the impedance

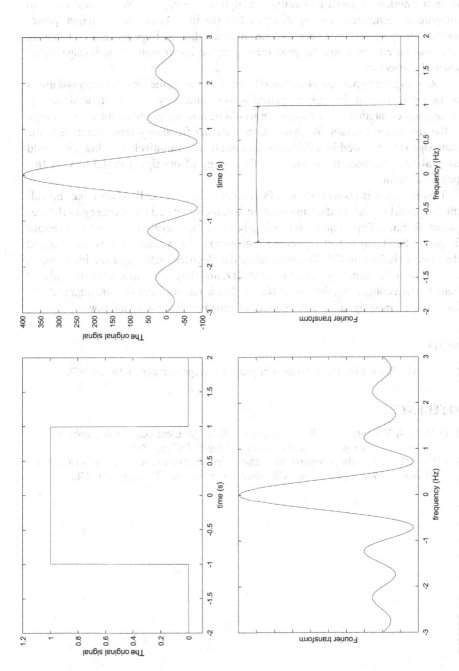

FIGURE 3.9   A pulse in the time domain will make a sinc in the frequency domain and vice versa

across which voltage is measured. I assume more readers of this book have read about the notion of power and voltage in high school physics. Now, if one wants to compute the amount of energy dissipated in the impedance, the computed power should simply be integrated across time. Imagine that the impedance is constant; in this case the energy would be proportional to the integration of the voltage signal squared across time.

Following the hypothetical scenario I outlined earlier, the signal energy is defined as the integration of the signal squared across time. Now that we understand frequency representation, a valid question would be how we calculate the signal energy in the frequency domain. We have stated that the frequency representation is the same signal manifested in a different domain; hence, intuitively speaking, it should not make a difference if we calculate the energy of the signal in the time or frequency domain.

This intuition is correct. Parseval's theorem mathematically shows that the calculated signal energy in the time domain is directly related to its energy in the frequency domain. This is quite interesting because the energy of a signal in a certain frequency band may reflect a specific phenomenon. For example, in the context of and electroencephalogram (EEG) representing the electrical activity of the brain, alpha waves have a frequency band of 8 to 12.99 Hz, and they are dominant in the wakeful state, but in a coma state, they are diffused (Rana, Ghouse, and Govindarajan, 2017). Thus, the average signal energy in a unit of time is called the signal power.

NOTE

1. These MATLAB files are available at https://www.crcpress.com/9780367207557.

REFERENCES

A. Q. Rana, A. T. Ghouse, and R. Govindarajan, "Basics of Electroencephalography (EEG)," in *Neurophysiology in Clinical Practice*, Springer, 2017, pp. 3–9.
J. Stillwell, "Logic and the philosophy of mathematics in the nineteenth century," in *Routledge History of Philosophy Volume VII*, Abingdon: Routledge, 2013, pp. 224–250.

4 Sampling of Continuous Time Signals

So far, I have discussed continuous time signals; however, these days, the use of computers in signal analysis is ubiquitous. Computers enable us to perform very heavy computational processes in a very short time. However, there is one issue here: no matter how fast a computer is, the temporal resolution of recoding and registering the signal is limited. We are simply unable to register the signals such that the signal amplitudes are retrievable for infinitesimal time steps. The questions is whether we really need this?

Throughout this chapter, I am going to explain why we do not need this unlimited temporal resolution; instead, we can work with a continuous time signal, given a limited temporal resolution if the continuous time signal is sampled with a fast enough rate. Because we work with a sequence of signal samples, the time axis is no longer continuous. Therefore, we call this sequence of samples a discrete time signal.

4.1 DISCRETE TIME SIGNALS

As mentioned earlier, discrete time signals are composed of a sequence of samples of a signal. One could imagine that they are the same as a continuous time signal sampled at equidistance intervals. Thus, for a discrete signal $x(n)$, one can write:

$$x(n) = x_c(n.T_s)$$

Where $x_c(t)$ is an arbitrary continuous time signal, but we have only its corresponding amplitudes at times that are integer multiples of T_s (the "sampling period"). To have an intuitive understanding of this, one could imagine that at time 0, we obtain the first sample of $x_c(t)$, then T_s seconds later we obtain the second sample, then T_s seconds later we obtain the third sample and so on. Mathematically speaking, this scenario, after some simplification, corresponds to the multiplication of $x_c(t)$ by a series of sharp impulses interspaced with T_s. Let us call this an impulse train $s(t)$. Note that $s(t)$ equals 0 at any time point except for times that are integer multiples of T_s, and at these time points its energy equals 1. I intentionally used the term "signal energy", as the signal amplitude for $s(t)$ at the impulse time is, mathematically speaking, undefined. The impulses are technically energy signals, and their amplitude is not defined in one single point. The multiplication of $s(t)$ and $x_c(t)$ will result in an impulse train whose energy at each of the impulses equals the amplitude of $x_c(t)$ at the corresponding time point. Figure 4.1 illustrates this process.

Thus, one can write:

$$x_s(t) = s(t).x_c(t)$$

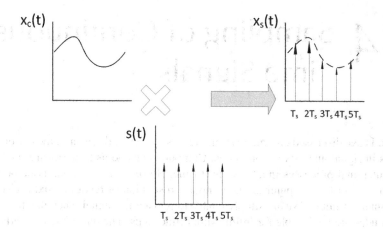

FIGURE 4.1 Multiplication of the impulse train $s(t)$ by a continuous time signal $x_c(t)$. Modified from Oppenheim, Willsky, and Nawab, 1997 and Oppenheim, Schafer, and Buck, 1989

Now, one can imagine that $x_s(t)$ equals 0 at any time point except for times that are integer multiples of T_s, and at these time points its "energy" equals $x_c(n.T_s)$. Thus, the energy of $x_s(t)$ consists of the sequence of samples defining $x(n)$, and if we just remove the time associated with the pulses of $x_s(t)$, we will actually obtain our discrete time signal.

4.2   SAMPLING IN THE FREQUENCY DOMAIN

The Fourier transform can also be defined for discrete time signals. Its mathematical formulation is quite similar to what I presented for continuous time signals:

$$x(t) \xrightarrow{\mathcal{F}} X(i\Omega) = \int_{-\infty}^{\infty} x(t)e^{-i.\Omega.t}dt$$

but for discrete time signals, one has to replace the integration with a regular summation. Thus, the Fourier transform for a discrete time signal $x(n)$ would be defined as:

$$X(e^{i\omega}) = \sum_{n=-\infty}^{\infty} x(n)e^{-i.\omega.n}$$

Now I have to show how the Fourier transform of a discrete time signal is related to the sampling of a continuous time signal. Note that I intentionally used a different notation to represent the angular frequency in the Fourier transform for a discrete time signal ω compared with Ω, which was used for continuous time signals. Furthermore, note that the angular frequency is represented by $i\Omega$ for continuous time signals, whereas it is represented by $e^{i\omega}$ for discrete time signals. This is not done by an accident—there is a good justification for this choice. This will be briefly explained when discussing systems and their response and stability in Section 7.6.

I mentioned before that any manipulation of signals in the time domain will be reflected in the frequency domain in some way. It can be mathematically verified that if a continuous time signal results from a multiplication of two signals in the time domain, its Fourier transform would correspond to the convolution of the Fourier transform of those two signals, and if the transform is presented in terms of angular frequency, the results would be scaled by $\frac{1}{2\pi}$ (Oppenheim, Willsky, and Nawab, 1997) (page 328). This nice property is called the convolution theorem of the Fourier transform. Appendix A.3 presents a short introduction to the concept of convolution. Interestingly, the Fourier transform of an impulse train in the time domain would also be an impulse train in the frequency domain. The following equation outlines this nice property of the Fourier transform of an impulse train (Oppenheim, Schafer, and Buck, 1989) (page 83):

$$\sum_{n=-\infty}^{\infty}\delta(t-n.T_s)\xrightarrow{\mathcal{F}}\frac{2\pi}{T_s}\sum_{k=-\infty}^{\infty}\delta(\Omega-k.\Omega_s)$$

Where $\Omega_s \frac{2\pi}{T_s}$ and $\delta(t)$ represent an impulse at time t, and $\delta(t-n.T_s)$ represents a shifted impulse by $n.T_s$.

The other interesting property of convolution is that the convolution of any signal by an impulse equals the same signal, and if a signal were convolved by a shifted impulse, the result would also be shifted to the same extent. Thus, the Fourier transform of $x_s(t)$ that I defined earlier would be the sum of the shifted versions of the Fourier transform of $x_c(t)$.

Figure 4.2 illustrates this point. Imagine that $x_c(t)$ has an arbitrary Fourier transform $X_c(i\Omega)$, shown in Figure 4.2a. $X_c(i\Omega)$ should satisfy certain conditions—among others, it has to have a limited frequency band, meaning that its amplitude should equal zero beyond a certain frequency range.

Now the Fourier transform of $x_s(t)$, $X_s(i\Omega)$ would look like something shown in Figure 4.2b. Based on the definition of a Fourier transform, one could verify that $X_s(i\Omega)$ can be expanded as shown:

$$X_s(i\Omega)=\sum_{n=-\infty}^{\infty}x_c(n.T_s)e^{-i.\Omega.T_s.n}$$

If one compares this equation and the definition of a Fourier transform for discrete time signals,

$$X(e^{i\omega})=\sum_{n=-\infty}^{\infty}x(n)e^{-i.\omega.n}$$

it is obvious that $X_s(i\Omega)$ is a lot like $X(e^{i\omega})$. If one simply replaces ω with $\Omega\,T_s$ in $X(e^{i\omega})$, $X_s(i\Omega)$ would be derived.

$$X_s(i\Omega)=X(e^{i.\Omega.T_s})$$

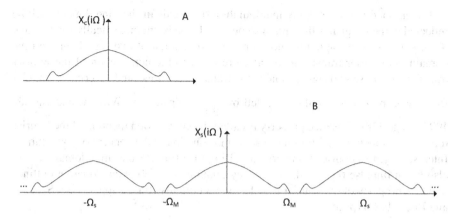

FIGURE 4.2 (a) An arbitrary Fourier transform, $X_c(i\Omega)$. (b) The summation of shifted versions of $X_c(i\Omega)$ representing the Fourier transform of $X_s(i\Omega)$, with the resulting signal from the multiplication of time representation of $x_c(t)$ and an impulse train with a period of T_s. Modified from Oppenheim, Willsky, and Nawab, 1997 and Oppenheim, Schafer, and Buck, 1989

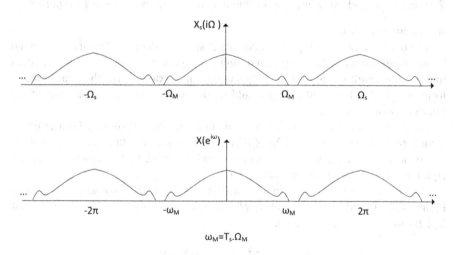

FIGURE 4.3 The relationship between the Fourier transform of continuous and discrete time signals after sampling. Modified from Oppenheim, Willsky, and Nawab, 1997 and Oppenheim, Schafer, and Buck, 1989

This is a nice property because based on this relationship, we can conclude that the Fourier transform for the discrete time signals is simply a scaled version of $X_s(i\Omega)$ in which $\Omega T_s = \omega$. This means that the frequency axis for the discrete time signal would simply be scaled such that $\Omega = \Omega_s$ would correspond to $\omega = 2\pi$. Figure 4.3 illustrates this frequency scaling. Another interesting observation in Figure 4.3 is that the Fourier transform of a discrete time signal is always periodic, with a period of 2π. Therefore, if we obtain the Fourier transform of a discrete time signal in one period of 2π, we know the frequency representation of the discrete time signal across the entire frequency domain.

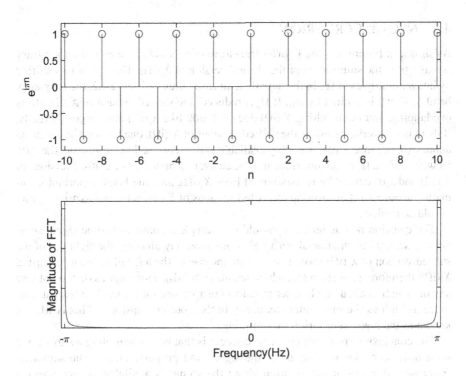

FIGURE 4.4   A signal with an intermittent change of the amplitude sign $e^{i.n.\pi}$

Very often, we choose to display the frequency axis in the range of $(-\pi,\pi)$. The lowest frequency component in this range would be zero and the highest frequency would be π. This is straightforward to imagine, because a discrete time signal with the highest frequency would be the one that intermittently changes the sign of its amplitude (Figure 4.4). Such a signal can be formulated as $x(n) = e^{i.n.\pi}$ whose Fourier transform would be the same as the Fourier transform of a constant signal $x(n) = 1$ but shifted by π in the frequency domain. The Fourier transform of $x(n) = 1$ is simply a discrete impulse train with a period of 2π starting from 0 radians per second. This would be the lowest imaginable frequency content, but when the signal has the highest frequency, its frequency representation is just shifted by π.

As can be seen in Figure 4.3 and the scaling relationship between the frequency domain for a continuous and discrete time signal $\Omega T_s = \omega$, the interval of $(-\pi,\pi)$ would correspond to $\left(-\dfrac{Fs}{2},\dfrac{Fs}{2}\right)$ in the frequency representation of a continuous time signal, where $Fs = \dfrac{1}{Ts}$ is the sampling frequency. In other words, when a continuous time signal is sampled with a frequency of Fs, the maximum frequency that can be represented in the frequency domain of the discrete time signal is $\dfrac{Fs}{2}$.

4.3   NYQUIST CRITERION

As shown in Figure 4.2, the Fourier transform of a sampled version of an arbitrary signal $x_c(t)$ is the summation of the shifted version of $X_c(i\Omega)$. The $X_c(i\Omega)$ was shifted by integer multiples of Ω_s. In the hypothetical case I drew in Figure 4.2, the frequency band of $X_c(i\Omega)$ is limited by Ω_M. If Ω_s is reduced, the shifted version of $X_c(i\Omega)$ starts overlapping, and the resulting $X_s(i\Omega)$ does not look like $X_c(i\Omega)$ any longer. Actually, if there is no overlap between the shifted versions of $X_c(i\Omega)$, one loses no information about $x_c(t)$, because $X_c(i\Omega)$ is readily available and can be used to reconstruct the $x_c(t)$. However, when Ω_s is too low, there will be an overlap between the shifted versions of $X_c(i\Omega)$ and $x_c(t)$ cannot be reconstructed from $X_s(i\Omega)$, and one loses important information about $x_c(t)$. This overlap of shifted versions of $X_c(i\Omega)$ is known as aliasing and should be avoided.

The question now is how fast should one sample a continuous time signal so as not to lose any information about $x_c(t)$? To not have any aliasing, the right tail of the shifted version of $X_c(i\Omega)$ should end before the rise of the left tail of the next shifted $X_c(i\Omega)$; therefore, $\Omega_s - \Omega_M > \Omega_M$, which means $\Omega_s > 2\Omega_M$. This implies that to not lose any information about the Fourier transform of $x_c(t)$, one must sample it with at least twice as high as the maximum frequency in the Fourier transform. This is what is known as the Nyquist criterion for sampling.

The conclusion from this essential concept is that before sampling a signal, one must have some knowledge about the signal and properly choose the sampling frequency. If no prior information about the signal is available, a conservative sampling frequency (quite high) should be chosen to ensure that no information is lost. For example, in case of a surface electromyography (EMG), we know that 95% of the signal power lies within the range of 5 to 400 Hz (Stegeman and Hermens, 2007). Before the sampling is conducted, a pre-filtering known as anti-aliasing is applied to limit the frequency band of the signal and reduce the aliasing effect. In case of an EMG, the cut-off frequency of this filter is set to 500 Hz; therefore, according to the Nyquist criterion, the sampling frequency should be at least 1000 Hz.

4.4   PRACTICAL CONSIDERATIONS

So far, I only explained the theoretical framework of sampling; however, in practice, this theoretical framework is too ideal to be directly implemented. For example, generating an impulse train is not practically feasible. Instead, at each sampling time, the amplitude of the signal can be stored and kept constant until the next sampling time. This is called a sample and hold. The entire procedure is equivalent to what is shown in Figure 4.5. Figure 4.5 shows that $x_s(t)$, which only exists in the theoretical framework, is applied to a system whose impulse response (its response to an impulse—this will be explained more in Chapter 7) is simply a pulse. The pulse width equals the sampling period, and its amplitude equals 1.

The other important point is about the analog-to-digital convertor (ADC), which was outlined in Chapter 2. The point is that when working with discrete time signals that are stored on a computer or an embedded system, not only is the time axis discretized but also the amplitude axis is quantized and essentially not continuous.

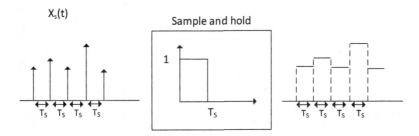

FIGURE 4.5 A block diagram showing an equivalent system for a sample and hold. Modified from Oppenheim, Willsky, and Nawab, 1997 and Oppenheim, Schafer, and Buck, 1989

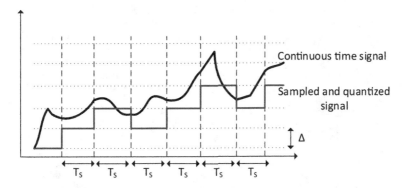

FIGURE 4.6 Illustration of sampling and quantization of a continuous time signal. T_s is the sampling period and Δ is the resolution of quantization

For example, for a surface EMG, the ADC resolution is recommended to be about 0.5 μV/bit, and therefore with such a resolution, a 16-bit ADC can cover a range of ±16 mV without considering the amplification gain (Stegeman and Hermens, 2007).

Figure 4.6 illustrates the discretization in time and the amplitude axis of a signal highlighting the previously mentioned points. It is clear that if the sampling frequency is not fast enough or the ADC resolution is quite coarse, the discrete signal deviates drastically from the original continuous time signal.

REFERENCES

A. V. Oppenheim, R. W. Schafer, and J. R. Buck, "Discrete-time signal processing," *Englewood Cliffs* (vol. 2). Prentice Hall, 1989.

A. V Oppenheim, A. S. Willsky, and S. H. Nawab, *Signals and Systems* (2nd ed.). NJ: Prentice Hall, 1997.

D. Stegeman and H. Hermens, "Standards for surface electromyography: The European project Surface EMG for non-invasive assessment of muscles (SENIAM)," *Enschede Roessingh Res. Dev.*, pp. 108–112, 2007.

5 Discrete Fourier Transform

In Chapter 4, I briefly introduced the Fourier transform of a discrete time signal:

$$X\left(e^{i\omega}\right) = \sum_{n=-\infty}^{\infty} x(n) e^{-i.\omega.n}$$

What this equation tells us is that as opposed to the time domain of a discrete time signal, which is itself discrete, its frequency representation is defined across a continuous scale. When one works with a sampled version of a signal, there is only a limited number of samples in a limited time window. Having another look at the previous equation, we realize that the time (n) should be expanded from minus infinity to infinity. How do we then calculate this equation for the samples outside our sampling interval? One assumption is to assume that the signal has zero amplitude outside of our sampling interval. According to what I outlined in Chapter 3, when the time representation of a signal has a limited non-zero domain across the time axis, its frequency representation will have an unlimited non-zero frequency domain. This is not a favorable property, because to perform sampling, the signal must have a limited frequency range such that Nyquist criterion can be satisfied and aliasing can be avoided.

Now imagine one samples the $X(e^{i\omega})$ in the frequency domain, which means that one gets to retrieve the amplitude of $X(e^{i\omega})$ only for a limited number of frequency points with equal intervals (frequency resolution). In accordance with the duality of the time and the frequency domains and in accordance with what I showed for the Fourier transform of a sampled signal in time, the time representation of the sampled $X(e^{i\omega})$ will be periodic in time.

That was a long sentence, so let me explain. We know that if we sample a signal in time, its Fourier transform is going to be periodic. Hence, according to the duality property, if one samples a Fourier transform in the frequency domain, one would then expect that this results in a periodic signal in time. This means that the sampling of the Fourier transform in the frequency domain would correspond to a signal in time with an unlimited time span; in turn, this means that such a signal would then have a limited frequency span in the frequency domain and we can work with it. Thus, instead of supposing that the signal with a limited number of samples has zero amplitude outside that range, we can assume that it repeats itself outside that range.

Imagine a discrete time signal $x(n)$, which has non-zero amplitude only within the range of 0 and $N-1$; if one sampled $X(e^{i\omega})$ with intervals of $\omega = \frac{2\pi}{N}k$ where $k = 0, \ldots, N-1$, the sampled $X(e^{i\omega})$ could be written as:

$$\tilde{X}(k) = X\left(e^{i\frac{2\pi}{N}k}\right) = \sum_{n=0}^{N-1} x(n) e^{-i\frac{2\pi}{N}k.n}$$

35

This is what is known as discrete Fourier transform (DFT). The previous equation is known as the "analysis equation", as it enables the frequency analysis of the sampled signals. It is straightforward to algebraically verify that the DFT is invertible, and the signal can be reconstructed based on its DFT.

$$x(n) = \frac{1}{N}\sum\nolimits_{k=0}^{N-1}\tilde{X}(k)e^{i.\frac{2\pi}{N}k.n}$$

This equation is called the "synthesis equation", as it enables the reconstruction of the time representation of the signal based on its DFT.

5.1   FREQUENCY RESOLUTION

As mentioned earlier, DFT is the sampled version of the Fourier transform of a discrete time signal. This sampling is done with steps of $\frac{2\pi}{N}$ in the frequency axis, where N is the number of signal samples in time. In other words, the frequency interval between two successive samples in the frequency domain is $\frac{2\pi}{N}$ radians per second. From Chapter 4, one remembers the scaling relationship between the frequency representation of a continuous time signal and a discrete time signal, $\Omega T_s = \omega$, where T_s is the sampling period and its reciprocal $F_s = \frac{1}{T_s}$ is the sampling frequency. Particularly, 2π radians per sec in the frequency domain of a discrete time signal corresponds to $2\pi.F_s$ in the frequency domain of a continuous time signal. Thus, the corresponding frequency resolution for the sampled continuous time signal is $\frac{F_s}{N}$.

Now imagine that a continuous time signal has a time span of T, which means that the signal has sampled within a time window between 0 and T sec of the signal with a sampling frequency of F_s. This will make $F_s.T$ (F_s times T) number of samples. In other words, $N = F_s.T$ for the discrete time signal, which results from the sampling of the continuous time signal. If one plugs N into the frequency resolution $\frac{F_s}{N}$, this results in an interesting finding: that is, the frequency resolution equals the reciprocal of the time span of the continuous time signal $\frac{1}{T}$.

This intuitively makes sense because if one needs to obtain a DFT that can reflect small changes in the frequency domain, the continuous time signal should be sampled across a longer time span in order for the small changes in the frequency domain to be accommodated. For example, if one thinks of a sine wave with a frequency of 1 Hz, a 1-sec time span would be required to reflect one full period of this sine wave, whereas this time span would 10 sec for a sine wave with a frequency of 0.1 Hz.

Based on this, we can conclude that if one needs to reflect changes of 0.1 Hz in the frequency axis, the continuous time signal should be sampled across a span of at least 10 sec, and if even finer resolution is required, a longer sequence of samples should be used in the DFT calculation. One may conclude then that the longer the time span of the signal, the better the frequency resolution. However, I will discuss this later and highlight that because of the "non-stationarity" of signals, we are not fully free to choose a very long time span, and this is apart from the technical issues that may arise when recording over a long period.

Earlier, I disregarded the fact that when the signal is analyzed in a limited time window, the frequency characteristics of the window itself contribute to the frequency resolution. This means that even though the frequency interval between two successive samples in the DFT is a small Δf, if the signal has two frequency components that are slightly further apart than the Δf, due to the frequency characteristics of the window, the DFT may still not be able to fully distinguish these two components from each other. I will get back to this point in Section 6.3.

5.2   FAST FOURIER TRANSFORM

An interesting property of DFT is that due to some of its computational features, some very efficient algorithms can implement the DFT algorithm much faster than implementing it directly based on its formulation. This approach of calculating DFT with a computationally efficient algorithm is known is fast Fourier transform (FFT). For example, a decimation-in-time algorithm (Oppenheim, Schafer, and Buck, 1989) (page 587 therein) is a well-known algorithm that benefits from the periodicity and symmetry of $e^{i \cdot \frac{2\pi}{N} k \cdot n}$ and reduces the number of multiplication and addition steps required to implement the DFT algorithm.

Particularly, the FFT algorithm is more efficient if the number of samples of the discrete signal N is a power of 2, for example, $N = 64, 128, 256, \ldots$. Therefore, even if the number of available samples is not a power of 2, the signal is padded by zeros ("zero padding") to fill in the rest of the samples with zeros and create a sequence with a power of 2.

5.3   ALIASING IN THE TIME DOMAIN

As I mentioned earlier, DFT is equivalent to the sampling of the Fourier transform, and this sampling in the frequency domain would correspond to a periodic signal in time, which is composed of copies of the original signal shifted by N. In Chapter 4, I discussed the overlapping of the shifted versions of the Fourier transform in the frequency domain and concluded that the extent of the shift should be such that the original Fourier transform can simply be reconstructed by isolating one of the shifted versions in one period. The same concept can be applied here, and the shifted versions of the signal in time should not overlap, and the original signal should be reconstructed by looking at the signal over one single period. The question here is that if the DFT algorithm is the sampling of the Fourier transform in the frequency domain, how many samples of the DFT are sufficient to be able to reconstruct the signal in time? It is straightforward to realize that the number of samples from the Fourier transfer should be greater than N so that the original signal can be reconstructed just by isolating one period of the time representation of the sampled signal in the frequency domain.

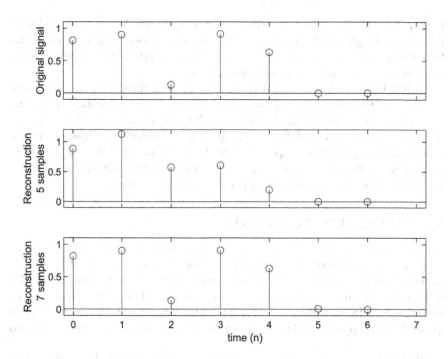

FIGURE 5.1   (a) an arbitrary signal with five samples; (b) the reconstruction of the corresponding time signal by sampling five samples from the Fourier transform; (c) the reconstruction of the corresponding time signal by sampling seven samples from the Fourier transform

Figure 5.1a shows a signal with a time span of five samples. Figure 5.1b shows the corresponding signal in time if the Fourier transform is sampled by only five samples and the time signal is reconstructed. Figure 5.1c shows the result of the same procedure when the Fourier transform is sampled by only seven samples. As can be seen, the reconstructed signal in Figure 5.1c is almost identical to the original signal, but the signal shown in Figure 5.1b differs quite considerably.

REFERENCE

A. V. Oppenheim, R. W. Schafer, and J. R. Buck, "Discrete-time signal processing," *Englewood Cliffs* (vol. 2). Prentice Hall, 1989.

6 Power Spectrum

Many biological signals have random properties. What does this mean? Imagine that one is capturing an electromyogram (EMG) showing the electrical activity of a muscle. Even if the experiment is fully controlled—for example, the level of contraction, the skin electrode impedance and the hardware for capturing the signal are kept consistent across the experiment—recording from the same muscle does not results in signals with completely identical samples if the recording is repeated. There will be many commonalities between the two signals indicating that the two signals represent the same level of muscle contraction, but if the signals are compared sample by sample, they will not be exactly the same.

Theoretically speaking, this scenario is a lot like the concept a random variable. What is a random variable? My intention is not to open a Pandora's box of stochastic processes and make this chapter more complex than it needs to be, but I will briefly describe what a random variable is. For example, when one tosses a coin or a die and registers the observations with each toss, the sequence of observations shows the values of a random variable. Obviously, before the values of a random variable are observed, it is associated with uncertainty. When one tosses a coin, one does not know what the outcome will be: Is it heads or tails? However, one may have an idea about the probability of the occurrence of the various possible outcomes. If one tosses a fair coin, the probability of observing a head or tail is 0.5.

This example pertains to a discrete random variable, which means that the outcome can be from a discrete set of events: the outcome of tossing a coin is either a tail or a head. However, one can also imagine a continuous random variable. For example, if a coin is being tossed on a surface of a table and we are interested in knowing the position of the point where the coin finally lands, we may use a coordinate frame to quantify the spatial location of the landing point on the surface of the table. Figure 6.1 illustrates such a hypothetical example.

In this case, the coordinate of the landing point changes on a continuous scale and is not limited to a discrete set of specific coordinates along the X- and Y-axis in Figure 6.1. Obviously, in this case, depending on the way the coin is being tossed and the location of the toss, it would be unlikely for the coin to land far from where one tosses it, but within the vicinity of that point, the toss is likely to land on any point. In the case of continuous random variables, we do not specify the probability of the occurrence of each point; rather, we specify the probability for certain intervals. We work the probability density function (PDF), and the probability of a random variable to be within a certain interval is obtained by calculating the area under the PDF in that interval. Biological signals such as an EMG can be regarded as a continuous random variable, and we may have a good idea about their corresponding PDF in some specific scenarios, such as the PDF of the EMG in a certain type and level of contraction. The concept of PDF can also be used with discrete random variables, and it is then called the probability mass function (PMF). When one reads something

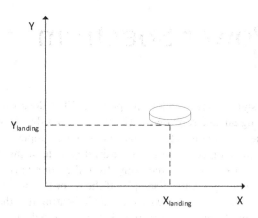

FIGURE 6.1 The final landing position of a coin on the surface of a table; a coordinate frame can be used to quantify the spatial location of the landing point on the table

about the distribution of a random variable, the distribution refers to the PDF or PMF for continuous or discrete random variables, respectively.

6.1 STATIONARITY

We have discussed the probable outcomes of a random variable. Now, let us add one more dimension to the picture: time. When we capture biological signals, we are actually observing the outcomes of a random variable across time. What role does time play here? If one keeps tossing a coin across time, would we expect that the probability of observing heads to change? Well, most likely not. However, imagine that the coin is made of a material that is not as rigid as metal. Then, with each toss of the coin, the coin deforms a little. If one keeps tossing that coin, after some time, it is not hard to imagine that the deformity of the coin causes it to land most likely with heads up or down. In that case, the probability of observing heads after each toss is not simply 0.5 any longer. In this case, one may say the distribution of the random variable is changing with time. Thus, if one were to estimate the PDF/PMF from the observations obtained at the beginning of the experiment, he or she would end up with a different estimate of the PDF/PMF compared to when the observations from the ending part of the experiment are used to estimate it.

The same notion of changing the statistical properties with time can be envisioned for signals such as biological signals. Figure 6.2 shows two signals across time. One of the signals preserves its statistical properties across time, but the other one shows some changes across time. The figure also shows the histogram of the signal (an empirical estimation of the PDF) at the beginning of the signal and at the end.

The random variable whose statistical properties are conserved across time is called a "stationary" random variable, and this property is called stationarity. Stationarity is quite a restricting assumption—most biological signals do not preserve stationarity. Therefore, a more relaxed alternative is introduced which is called "stationarity in a wide sense". A random variable with a wide-sense stationarity does

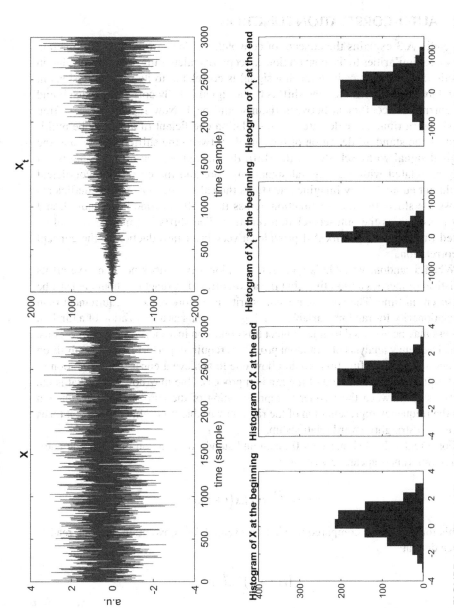

FIGURE 6.2 Two arbitrary signals, one with consistent statistical properties across time (AR process) on the left and one with variable coefficients across time on the right. The histogram of the signal at the beginning and the end of the signal shows that the distribution of the signal with variable coefficients clearly changes.

not need to conserve all its statistical properties, but the auto-correlation function should at least be consistent in time.

6.2   AUTO-CORRELATION FUNCTION

Appendix A.3 explains the concept of convolution. The auto-correlation function has some similarities to the convolution concept. By calculating the auto-correlation function, one quantifies how much a signal is correlated to its shifted versions in time. Obviously, when the time shift is 0, the signal is fully correlated to itself and the correlation coefficient between them would be 1. Now if one starts shifting the signal in time and calculates the correlation coefficient (it also may be multiplied by the standard deviation of the signal) between the shifted version and the original signal with each step of the shift, the shifted version of the signal is not fully correlated with the original signal anymore and the correlation coefficient would not equal 1. Now imagine one stores the calculated correlation coefficients across the shifts in time—a function across the shifts in time is composed, and that is known as the auto-correlation function. The shifts in time are very often called time lags. Appendix A.4 provides a very brief introduction to the concept of correlation.

When a random variable is wide-sense stationarity, it does not conserve all its statistical properties across time, but the auto-correlation function should at least be consistent in time. The wide-sense stationarity is a more relaxed requirement than the stationarity for random variables. Biological signals as a realization of a random process may be assumed to at least meet this condition in a short time window. Thus, if the frequency analysis of a random process is required, performing the analysis on the auto-correlation function would allow one to achieve a consistent estimation of the frequency representation of the random process. One may now ask: What is the relationship between the frequency representation of the auto-correlation function and the frequency representation of the random variable itself? Interestingly, it turns out to be a straightforward relationship.

For a real valued signal with 0 mean and standard deviation of 1 $x(t)$, the auto-correlation function can be written as:

$$r_x(t) = \int_{-\infty}^{\infty} x(\tau).x(t+\tau)d\tau$$

If this formulation is compared to what I presented for convolution, one can readily conclude that:

$$r_x(t) = x(t) * x(-t)$$

This equation help us understand the frequency representation of the auto-correlation function. Remembering the convolution theorem (see Section 4.2), one can write:

$$R_x(j\Omega) = X(j\Omega).X(-j\Omega)$$

Where $X(j\Omega)$ is the Fourier transform of $x(t)$; note that $x(t)$ is a real valued signal, meaning that the signal amplitude is a real number and not a complex number. A nice property for a real valued signal is that we can write:

$$X(-j\Omega) = X*(j\Omega)$$

Where $X*(j\Omega)$ is the conjugate of $X(j\Omega)$. To remember what conjugate means, see Appendix A.2.

Now, one can realize an interesting relationship between the Fourier transform of the original signal and its auto-correlation function:

$$R_x(j\Omega) = |X(j\Omega)|^2$$

Where $|X(-j\Omega)|$ is the magnitude of the Fourier transform of $x(t)$. A similar concept can be developed for discrete time signals, and that would be more relevant to our problems because we often work with discrete time signals.

In the previous calculation, we have one more assumption, and that is called "ergodicity", which means that the statistical properties of a random process can be obtained from the timeline of its realizations.

6.3   SPECTRAL ESTIMATION

In Section 6.2, I outlined the definition of the auto-correlation function of a signal and showed that it is in itself a signal in time, and its frequency representation has a straightforward relationship to the frequency response of the signal. The Fourier transform of the auto-correlation function is called PSD, as this represents how the signal power is distributed along the frequency axis.

Various algorithms can be used to obtain an estimation of PSD. Generally, they are divided into three main categories: non-parametric methods, parametric methods and subspace methods. The latter is not within the scope of this book, but I will outline some aspects of non-parametric and parametric methods to obtain a general understanding of them.

6.3.1   Non-Parametric Methods

At first glance, this concept may seem quite trivial. One estimates the auto-correlation function and then takes the FFT and that is it—now we have an estimation of the PSD. Even though the essence of the work may be as simple as that, technical challenges can be encountered. For example, the effect of noise and windowing are among the issues that should be investigated. Another important point is the concept of bias and variance of estimation.

What is the bias of estimation? Our mathematical formulation for an auto-correlation function needs to be worked out in the range of $(-\infty,\infty)$. Obviously, we do not get to do this calculation, as we only work with a limited number of samples. We expect that as the number of samples involved in the calculations increases, the mean of estimation gets close to the mean of what we want to esti-mate, which in this case is the PSD.

However, it is not always guaranteed that the mean of our estimation tends to the mean of what we want to estimate. The difference between the mean of estima-tion and the mean of what we want to obtain is called the bias of estimation. With respect to the variance of the estimation, we expect that as we increase the number of samples in the estimation, the estimated values become less spread out around the mean of the estimation. This property is called the consistency of the estimation, but not all estimators are consistent.

To perform the calculations, we can either take the FFT of the signal and then calculate the square of its magnitude (periodogram) or estimate the auto-correlation function and then take its FFT (correlogram) (Poularikas, 2017) (page 214). The periodogram asymptotically converges to PSD and has no bias, as the number of samples tend to infinity, but it is not a consistent estimator of PSD. The estimation of the auto-correlation function can be biased or unbiased, but the biased estimator is often used in the calculations, as it has some computational advantages. A discussion of the advantages is beyond the scope of this book.

Figure 6.3 shows the periodogram of a synthetically generated signal. The signal is made of two specific frequency components at 100 and 102 Hz, as well as some random noise.

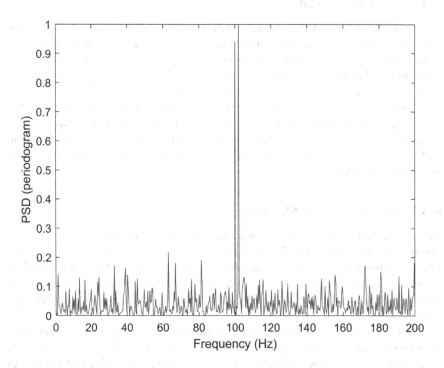

FIGURE 6.3   The periodogram of a synthetic signal. The two frequency components at 100 and 102 Hz are clearly obvious compared to the background noise.

The periodogram looks very jaggy in appearance, as the variance of estimation is quite high. As mentioned earlier, a periodogram is not a consistent estimator of PSD; thus, even increasing the number of samples does not reduce the estimation variance considerably. Further, note that the frequency range in displaying the PSD includes only the positive frequency range. This is often called a one-sided PSD. If a two-sided PSD is invoked, the PSD will be displayed over a full period of frequency representation between zero and the sampling frequency.

One way to reduce this relatively high estimation variance is to apply an averaging filter on the estimated PSD (Daniel periodogram) (Manilo and Nemirko, 2016). Other algorithms have also been developed to reduce the estimation variance in periodograms. A very commonly used algorithm is called the Welch method. The Welch method splits the signal into smaller overlapping epochs and then estimates the PSD for each of the epochs. Finally, the overall estimation of PSD is obtained by averaging the estimated PSD from each of the epochs.

Note that the Welch method assumes the stationarity of the signal in the given time window; thus, the estimated PSD in each of the epochs is assumed to be derived for realizations of the same random variable, and therefore the averaging is expected to reduce the estimation variance. Figure 6.4 shows the Welch PSD of the same synthetic signal used in Figure 6.3.

Comparing Figure 6.3 and Figure 6.4 clearly shows that the Welch PSD looks much smoother—in other words, the estimation variance has been reduced considerably.

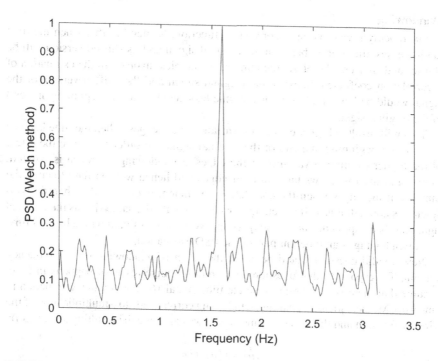

FIGURE 6.4 The Welch power spectrum of a synthetic signal. The signal is the same as the one used in creating Figure 6.3

However, the two frequency components at 100 and 102 Hz, which were clearly distinguishable in Figure 6.3, are now smudged into one frequency component. Obviously, when the original signal is divided into smaller epochs, the frequency resolution will not remain as fine. This is why the two frequency components at 100 and 102 Hz are not distinguishable any longer. The bottom line is that one has to pay a price to reduce the estimation variance, and that price in this case is coarsening of the frequency resolution. Nothing comes for free.

Other ways of splitting the signal into smaller epochs have been suggested to improve the frequency resolution in the Welch method while the estimation variance still remains low. For example, the epoch could be chosen with unequal length in the asymmetric modified Welch and symmetric modified Welch methods (Poularikas, 2017) (Page 236). However, these algorithms provide an estimation with properties between the periodogram and the Welch methods. This means that the frequency resolution is coarser than the periodogram but finer than the Welch method, and the estimation variance is lower than the periodogram but higher than the Welch method. The general rule still holds: nothing comes for free!

When estimating the PSD, often one has to choose a proper window to control the frequency leakage (see the next section on windowing) and resolution. Additionally, depending on the number of samples involved in the FFT calculation, the signal may need to be padded by zeros in the end. Next, I briefly describe what windowing and zero padding are.

Windowing

Let us quickly revisit the auto-correlation function. Remember that when the time lag increases, the overlap between the original signal and its shifted version will be shorter and shorter. Therefore, the number of samples involved in the estimation of a correlation coefficient between the original signal and the shifted version of the signal would be lower and lower as the time lags get close to the beginning and end of the original signal.

This will result in higher estimation variance at the edges, when the time lag gets close to the beginning and end of the original signal. In order to reduce the effect of the higher estimation variance at the edges, a windowing approach is utilized. Commonly used windows taper at their edges and hence weaken the effect of the samples at the edges when the correlation coefficient is estimated. When no windowing is applied, it is as if a rectangular window with the same size as the original signal has been applied. Thus, as long as one works with a signal with a limited time span, windowing is an inherent part of the PSD estimation.

Now a valid question would be about the effect of windowing in the frequency domain. To respond to that question let us remind ourselves what happens in time. Assume that there is an arbitrary discrete time signal $x(n)$ and a window function in time $w(n)$. Mathematically speaking, windowing corresponds to a multiplication of the window function and the signal in time. Thus, we can write the resulting signal $x_r(n)$:

$$x_r(n) = x(n).w(n)$$

Now it is easier to say something about the frequency. For discrete time signals, the multiplication in time corresponds to an operation in the frequency domain, which is called "periodic convolution". Do not be scared of the this term—it is quite similar to the convolution described in Appendix A.3. However, bear in mind that the Fourier transform of a discrete time signal is a periodic function, so one has to calculate the convolution integral of two periodic signals with the main period of 2π. The Fourier transform of $x_r(n)$ is periodic itself, thus the periodic convolution should also be calculated in a period of 2π. Let us show the periodic convolution with $\otimes$ just to discriminate it from the regular convolution denoted by *.

$$X_r\left(e^{i\omega}\right) = X\left(e^{i\omega}\right) \otimes W\left(e^{i\omega}\right)$$

Where $X(e^{i\omega})$ and $W(e^{i\omega})$ are the Fourier transforms of $x(n)$ and $w(n)$, respectively. Note that a somewhat similar concept is defined for the DFT, and it is called circular convolution. If two discrete time signals are multiplied in time, the DFT of the product is obtained by the circular convolution of their DFTs. This will not be discussed further in this chapter.

Let us imagine that $w(n)$ is constant across the entire time axis without any limits. The $W(e^{i\omega})$ in that case would be an impulse train with a period of 2π and $X_r(e^{i\omega}) = X(e^{i\omega})$. But we do not work with a signal with no limit in time—our analysis is done in a limited time window. Now imagine we just have a limited number of samples and we do not apply any particular windowing, which is equivalent to applying a rectangular window as mentioned earlier.

Let us see what the effect of windowing would be in the frequency domain. For starters, I describe the case for a rectangular window. Figure 6.5 shows the magnitude of the Fourier transform of a rectangular window. As can be seen, the energy of the signal is condensed in certain frequency ranges called lobes. The biggest lobe, which is centered on 0 Hz, is the called the main lobe, and next to the main lobe, the side lobes are seen.

As the length of the window increases, the width of the lobes decreases, but the area under the curve remains constant. It is straightforward to imagine that the oscillations in the frequency representation of the window $W(e^{i\omega})$ will have an effect on the frequency representation of the windowed signal $X_r(e^{i\omega})$.

One of the very important results of windowing is something called frequency leakage. Figure 6.6 depicts a signal that has two frequency components at 50 and 60 Hz. The two components have the same energy level; however, when a rectangular window is applied to the signal, the frequency representation of the signal detects two components with different energy levels. This is because a part of the signal energy in the existing frequency components has leaked to the other components. This phenomenon is primarily affected by the relative size of the main lobe to the side lobe—the greater the amplitude of the side lobes with respect to the main lobe, the more influential the leaking effect. A rectangular window has relatively large side lobes, and this leads to a considerable leakage effect. Hence, a rectangular window is not very popular in estimating PSD.

There are quite a few alternative windows applicable in this context; for example, "Hamming" and "Hanning" windows are commonly used in PSD estimation. These windows taper at their edges and therefore have smaller side lobes compared to a rectangular window with the same length. In Figure 6.6, one can see that when the signal is windowed by a Hanning window, the energy of the two frequency

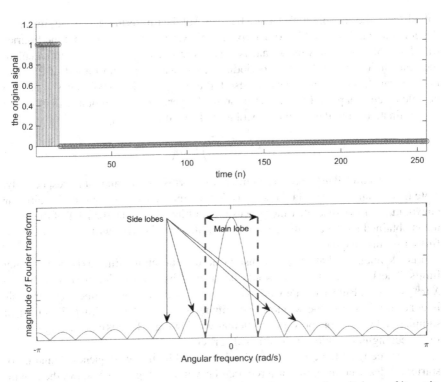

FIGURE 6.5 The magnitude of the Fourier transform of a rectangular window and its main and side lobes

components are represented more evenly compared to a rectangular window. Note that this does not completely eradicate the leakage effect but it does reduce it. This comes at a price of compromising the frequency resolution, however. The width of the main lobe in the Hanning window is wider than that of the rectangular window, and as one can see in Figure 6.6, the rectangular window is more effective for separating the two frequency components in the signal compared with the Hanning window. The leakage effect is related to a property of a window that is often called the "dynamic range", which refers to the ability to discriminate frequency components with different energy levels.

The Kaiser window is a parametrized window that allows one to choose a structural parameter called β which determines the relative size of the side lobe with respect to the main lobe (Oppenheim, Schafer, and Buck, 1989) (Page 452). Similar to other windows, the length of the Kaiser window will then determine the width of the main lobe and therefore the frequency resolution.

Now that I have discussed the characteristics of windows, I should point out what I mentioned about the frequency resolution in Chapter 5. There, I stated that because we work with a limited number of samples of a signal, the frequency characteristics of the window will have an effect on the frequency resolution. This is determined by an index that is called the equivalent noise bandwidth (ENBW). ENBW is the bandwidth (see Chapter 7) of a noise whose spectrum looks like a rectangle, and its total energy

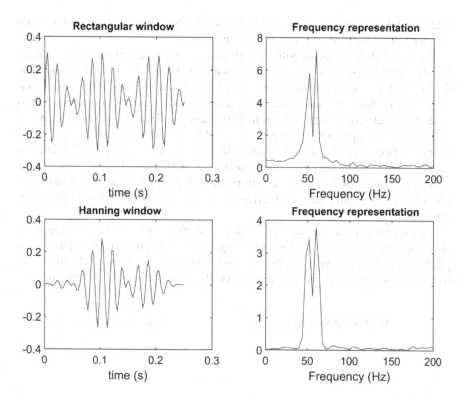

FIGURE 6.6 Effect of window type on frequency resolution and leakage

is equal that of the window. With this definition, a rectangular window has the lowest ENBW compared to other windows with the same length. This once again indicates that the rectangular window will have a finer resolution compared to other windows.

Zero Padding

The base of the non-parametric estimation of PSD is the calculation of FFT. As I mentioned before, the algorithms of an FFT will run faster and are computationally more effective if the signal length is a power of 2. Now if the actual signal does not have a length of a power of 2, one can add some zeros at the end of the sample sequence to make its length a power of 2. The question now is how this procedure affects the frequency analysis. Think back to frequency resolution. Essentially, zero padding is equivalent to having a signal with a longer length than the original signal, which is then multiplied by a rectangular window with the same length as the original signal. Therefore, what I mentioned about windowing is relevant here too. However, there is one point here to highlight. Let us assume that the FFT is one-sided and shows the positive frequencies. It spans a frequency interval between zero and $\frac{F_s}{2}$, and as one increases the number of samples involved in the FFT algorithm, the frequency interval between successive samples in the FFT are going to be shorter.

Then a valid question could be whether this means that the frequency resolution has become finer if one pads the signal with some zeros.

Figure 6.7 shows a signal with two adjacent frequency components at 1 and 1.5 Hz. The original signal has been sampled with 400 Hz for 750 ms. As can be seen, the FFT without zero padding cannot show the existence of these two components. Now imagine that one padded the sampled signal and recalculated the FFT. In this case, the signal was zero-padded and had 2048 samples. If the sampling frequency is 400 Hz, 2048 samples should correspond to 5.1 sec of the signal, and the frequency interval between successive FFT samples should be 0.2 Hz. If we had such a fine frequency resolution, the frequency components of 1 and 1.5 would have been identified in the FFT. However, as can be seen in Figure 6.7, the FFT estimation looks much smoother, but it is still unable to discriminate between those two adjacent frequency components.

This seems quite trivial, as zero padding does not bring new information about the signal; thus, if we do not have the frequency resolution in the first place, zero padding does not improve the frequency resolution. However, note that the frequency interval between two successive samples on the FFT of the zero-padded signal is

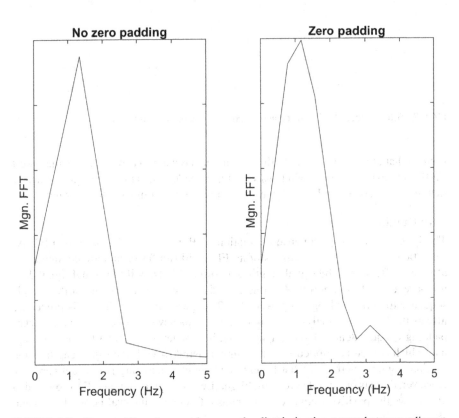

FIGURE 6.7 Zero padding does not increase the discrimination power between adjacent frequency components

indeed shorter. Thus, the FFT should be calculated for denser points along the frequency axis. This is essentially the results of the circular convolution of the original signal FFT and the rectangular window that I mentioned initially. One can imagine that the zero-padding procedure is somewhat similar to an interpolation in the frequency domain, and the FFT may be interpolated between two successive samples of the FFT of the original signal, but the interpolation would not be a simple linear interpolation.

6.3.2 PARAMETRIC METHODS

When PSD is estimated by means of non-parametric methods, we assume we do not have much information about the underlying phenomenon that generated the signal—we just have the regular assumptions that we had for the FFT calculation. However, in certain scenarios, we may be able to model the signal and then describe the signal variation as a limited set of parameters, which are the model parameters. That is why these methods are called parametric. We may achieve a much better frequency resolution with these methods. However, these methods are only accurate if we are able to construct a model that can sufficiently describe the variation in the signal and we can develop computational methods to estimate the model parameters based on the samples of the signals. As in the case of non-parametric methods, stationarity is a key presumption for these methods too. To better understand this concept, one should first be familiar with the concept of a system, which will be introduced in Chapter 7. Additionally, to fully comprehend how they work, understanding mathematical foundation of the methods is also essential. In Chapter 7, I will outline the essential concepts of parametric methods without going too much into the mathematical details.

6.4 SIGNAL CHARACTERISTICS BASED ON THE POWER SPECTRUM

Let us assume that we have an estimate of PSD at hand: How do we boil it down to some metrics so as to compare two different signals? For example, in studying an EMG, it is commonly known that the development of fatigue will induce some spectral changes in the EMG signal (Merletti, Lo Conte, and Orizio, 1991). Very often, the central tendency of PSD is of note. Indices such as mean power frequency (MPF), sometimes also abbreviated as MNF, and median power frequency (MDF) are broadly used in the literature to describe spectral changes of the EMG signal. Often, it is reported that due to increased synchronization between motor units and a reduction in the conduction velocity of motor unit action potentials across the muscle fibers, fatigue will shift the PSD to a lower frequency range, and therefore the central tendency indices (e.g., MNF and MDF) are reduced as fatigue develops. Figure 6.8 shows an example of such a case where MNF and MDF have dropped during a prolonged session of repetitive shoulder movements.

Some studies report zero crossing of a signal, which basically refers to the number of times a signal crosses the zero line in a given period, assuming that the signal has

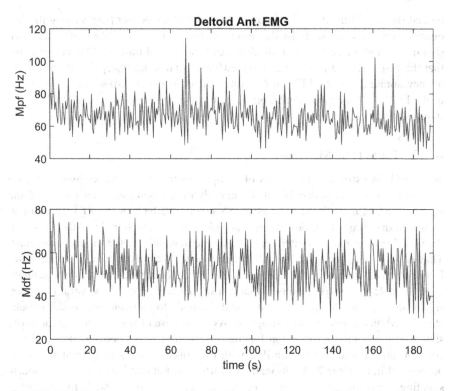

FIGURE 6.8   Mean power frequency and median power frequency during a repetitive task

a zero mean. To calculate the zero-crossing rate, one does not need to estimate the PSD, but zero crossing has close ties to the central tendency of PSD. For example, an increased rate of zero crossing is expected to be associated with a higher central tendency of the PSD. However, estimation of the PSD provides much more information. For example, in the analysis of an electroencephalogram (EEG), the brain activity is analyzed in certain frequency bands (e.g., α [8 to 13] Hz, β [>13 Hz], γ [4 to 8]Hz) with specific definitions of their frequency ranges. The variation of the signal power across these certain frequency bands may convey insightful information about the mental state and cognitive processing (Teplan, 2002). For example, the α wave power is expected to be dominant during a relaxed mental state with the eyes closed, and it is expected to be reduced when the individual feels drowsy or sleepy (Rana, Ghouse, and Govindarajan, 2017).

REFERENCES

L. A. Manilo and A. P. Nemirko, "Recognition of biomedical signals based on their spectral description data analysis," *Pattern Recognit. Image Anal.*, vol. 26, no. 4, pp. 782–788, 2016.

R. Merletti, L. R. Lo Conte, and C. Orizio, "Indices of muscle fatigue," *J. Electromyogr. Kinesiol.*, vol. 1, no. 1, pp. 20–33, 1991.

A. V. Oppenheim, R. W. Schafer, and J. R. Buck, "Discrete-time signal processing," *Englewood Cliffs* (vol. 2). Prentice Hall, 1989.

A. D. Poularikas, *Understanding Digital Signal Processing with MATLAB®and Solutions.* CRC Press, 2017.

A. Q. Rana, A. T. Ghouse, and R. Govindarajan, "Basics of Electroencephalography (EEG)," in *Neurophysiology in Clinical Practice*, Springer, 2017, pp. 3–9.

M. Teplan, "Fundamental of EEG Measurement," *Meas. Sci. Rev.*, vol. 2, no. 2, pp. 1–11, 2002.

7 Systems and Their Properties

Up to this point, I wrote about signals and their properties, but the essential goal when analyzing signals is to find out some information about the underlying mechanisms that generate them. Let us say we are given a black box and we want to understand what that black box does and perhaps how it does it. How do we find that out? We often turn the box around and notice how it moves—maybe we listen to the noises that it makes when moving. If we just describe what we do based on our common sense in the context of signal processing, we would say that we stimulate the black box with various inputs (for example, mechanical forces to move and turn it around) and then study the response (output) that it generates, for example, its velocity and acceleration profile.

This is a lot like systems in signal processing. A system can receive one (or more) signal in the input and then generate a response, which is observable as one (or more) signals. Studying systems is about finding the relationship between the input and output signals.

Just as signals were categorized into continuous and discrete time signals, systems can be categorized into discrete and continuous time systems. Continuous time signals can be electrical, mechanical or chemical systems that we work with on a daily basis. Very often, the electrical equipment that we work with is composed of multiple subsystems, including filters, amplifiers and so on. Discrete time systems are used in digital signal processing, and they work on discrete time signals. Very often, they are embodied in form of a computer algorithm or piece of code implemented in an embedded system, like a microcontroller. Here, we deal with the systems as an abstract concept that determines the relationship between input and output.

To denote the relationship between the input and output of a system S, let us assume that $x(t)$ is the input to the system and $y(t)$ is the output. For a discrete time system, we denote them by $x[n]$ and $y[n]$. We would then write:

$y(t) = S\{x(t)\}$ A continuous time system S and $y[n] = S\{x[n]\}$.

Figure 7.1 shows how we usually depict the system and its input and output.

FIGURE 7.1 Illustration block diagram of a system

7.1 SYSTEM PROPERTIES

Certain properties of systems are essential for understanding the relationship of their inputs and outputs and the implementation of the systems. In this section, I briefly introduce these properties.

7.1.1 LINEARITY

This property is probably the most common term, even among people with no technical background. You may have heard terms like linear system, linear equation, linear relationship, etc. The first thing that pops into mind is that perhaps the relationship between input and output is defined by a line. This is not quite far off, but there is a more general and useful definition of the concept. This property constraints the relationship between the input and output of a system to follow certain rules. Let us assume that the response of a system S to an input of $x_1(t)$ is $y_1(t)$, and the system response to $x_2(t)$ is $y_2(t)$; also suppose that α_1 and α_2 are constants. Then, if S is linear, its response to $\alpha_1 x_1(t) + \alpha_2 x_2(t)$ will be the same linear summation of the responses, which means:

$$S\{\alpha_1 x_1(t) + \alpha_2 x_2(t)\} = \alpha_1 y_1(t) + \alpha_2 y_2(t).$$

Any system that satisfies this relationship between input and output is a linear system; otherwise, the system is called non-linear. This relationship is a combination of two principles, namely superposition and homogeneity. Superposition implies that the response of a system to a summation of inputs is simply the summation of the system's responses to each of the inputs alone, which is:

$$S\{x_1(t) + x_2(t)\} = y_1(t) + y_2(t)$$

Note that it is not necessary to have only a summation of two signals in the input—it can be a summation of multiple inputs and, in turn, the output will also be the summation of the responses to each of the inputs.

Homogeneity implies that the response of a system to an input multiplied by a constant is the same constant multiplied by the response of the system to that input, which is:

$$S\{\alpha_1 x_1(t)\} = \alpha_1 y_1(t)$$

Thus, a system that satisfies superposition and homogeneity is a linear system.

Note that superposition and homogeneity are not simply two forms of the same concept. One can think of a system that only satisfies one of them. For example, take a system S denoted by the following relationship between input and output:

$$S\{x_1(t)\} = Re\{x_1(t)\}$$

Where *Re* represents the real part of a complex number. Such a system satisfies the superposition but not the homogeneity. The real part of the summation of two complex numbers is simply the sum of their real parts, but in cases where a constant whose imaginary part is not zero is multiplied by the input, the real part of the product is not the same as the multiplication of that constant and the real part of the input. The exact same definition can be generalized to discrete time systems.

Another interesting example is the Fourier transform itself. If the Fourier transform of $x_1(t)$ and $x_2(t)$ are $X_1(i\Omega)$ and $X_2(i\Omega)$, respectively,

$$x_1(t) \xrightarrow{\mathcal{F}} X_1(i\Omega)$$
$$x_2(t) \xrightarrow{\mathcal{F}} X_2(i\Omega)$$

Then the Fourier transform of $\alpha_1 x_1(t) + \alpha_2 x_2(t)$ will be $\alpha_1 X_1(i\Omega) + \alpha_2 X_2(i\Omega)$.

$$\alpha_1 x_1(t) + \alpha_2 x_2(t) \xrightarrow{\mathcal{F}} \alpha_1 X_1(i\Omega) + \alpha_2 X_2(i\Omega)$$

This means that the Fourier transform is a linear procedure. Please note that this is about the Fourier transform itself and not about its magnitude and phase, meaning that the magnitude of the Fourier transform of a summation does not equal the summation of its magnitudes.

Linearity is a nice property of a system, as the assumption of linearity simplifies system analysis. One of the nice features of linear systems is that if the input to a linear system consists of only a few monochromatic frequency components (e.g., a weighted sum of a few sine waves), the response of the linear system to that input can only consist of the same frequency components. Of course, the weighting of those components can vary, depending on the system properties. Using this property, Farina and his colleagues have shown that the common motor neuron pool responds linearly to the cortical and peripheral oscillatory inputs (Farina, Negro, and Jiang, 2013).

7.1.2 TIME INVARIANCE

Time invariance refers to whether the response of a system is changing according to time. To formally state the relationship between input and output, one could write that if the response of a system S to the input of $x_1(t)$ is $y_1(t)$, the response of S to a shifted version of $x_1(t)$ by τ will be the shifted version of $y_1(t)$ by the same time shift of τ, which is

$$S\{x_1(t-\tau)\} = y_1(t-\tau)$$

Or for discrete time systems where the shift in time is denoted by k:

$$S\{x_1[n-k]\} = y_1[n-k]$$

This relationship simply implies that if a system is time invariant, no matter when one stimulates the system with a specific input, the system response is going to be same, except that the time line of the input and output is simply shifted by the time that the input stimulates the system.

Now imagine a system like the following:

$$S\{x_1(t)\} = t.\,x_1(t)$$

It is quite straightforward to show that this system is linear, but it is not time invariant, as the multiplier (t) is changing with time. This is like an amplifier whose gain increases as time goes by.

7.1.3 CAUSALITY

This property implies that the output of the system does not depend on its future inputs. At first glance, this seems trivial. This is because systems in the analog world are all causal and it is hard for us to imagine a system whose response depends on its future inputs. However, in digital processing, it is quite normal to do of-line processing and analyze a batch of data. In that case, it is quite normal to imagine that at each point in time, the output of the system depends on the past and future samples of the input. We will see an example of a non-causal system when introducing bidirectional filtering.

7.1.4 STABILITY

This property refers to the boundedness of the output. If an input to the system is bounded—which means that the amplitude of the input signal is never above a constant upper limit and never below a constant lower limit—the output of the system is bounded too. This may seem quite trivial too, but I assume many readers of this book have seen an unstable system. Imagine that you are working with an electrical appliance and suddenly you realize that it starts burning out, and you may even see smoke coming out of the device. In such cases, you are observing an unstable system that does not function properly anymore. The collapse of the Tacoma Narrows Bridge in the United States is a classic case of an unstable system in which the construction of the bridge as a mechanical system responded recklessly to the wind (Billah and Scanlan, 1991).

7.2 LINEAR TIME-INVARIANT SYSTEMS

A system that satisfies both linearity and time invariance is called a linear time-invariant (LTI) system. In reality, both linearity and time invariance are quite restrictive, and it is almost impossible to find a system in practice that fully satisfies these conditions; however, some systems follow the rules somewhat closely, and it is convenient for us to assume these properties to facilitate studying those systems.

A very important corollary from the assumption of LTI is that if the response of an LTI system to an impulse is known, one can calculate the response of the system to almost any given $x(t)$ in the input. In previous chapters (e.g., Chapter 4), I wrote about an impulse that in mathematical terms is represented by a Dirac delta $\delta(t)$, which is a function whose total energy is condensed into one single time point (zero) and its amplitude is therefore zero everywhere other than at zero. Note that mathematically speaking, the amplitude of $\delta(t)$ at zero is undefined and tends to infinity. The response of the system to an impulse is called the impulse response and very often is denoted by $h(t)$.

The response of an LTI system with an impulse response $h(t)$ to an arbitrary input $x(t)$ ($x(t)$ has to meet certain conditions) is calculated by a convolution of $h(t)$ and $x(t)$. Chapter 4 introduced the concept of convolution.

$$y(t) = h(t) * x(t)$$

This is interesting because if we look at the relationship between input and output in the frequency domain, we can write:

$$Y(i\Omega) = H(i\Omega).X(i\Omega)$$

In other words, the ratio between the Fourier transform of the output $Y(j\Omega)$ and the input $X(j\Omega)$ determines the Fourier transform of the impulse response.

$$H(i\Omega) = \frac{Y(i\Omega)}{X(i\Omega)}$$

This is a unique function that determines how an LTI system responds to various inputs; therefore, it is called a "transfer function" or a "system function". If the system input is simply a direct delta (i.e., an impulse), $X(i\Omega) = 1$, then $H(i\Omega) = Y(i\Omega)$, which means that the transfer function is the Fourier transform of the system impulse response in the frequency domain.

A similar relationship can be derived for discrete time systems. However, an impulse in discrete time is not a Dirac delta anymore, but simply a function whose amplitude is zero everywhere, but it equals one at the zero time point. The convolution is also calculated for discrete time signals. The transfer function of a discrete time system can still be written as:

$$H(e^{i\omega}) = \frac{Y(e^{i\omega})}{X(e^{i\omega})}$$

As in the case of continuous time systems, the transfer function is the Fourier transform of the system impulse response for discrete time systems in the frequency domain. The transfer function is often expressed in terms of the Laplace transform, as it extends the concepts of the Fourier transform. Not only does the Laplace transform represent the frequency of oscillations in the system response, but it also concerns either the decay or buildup of the oscillations in the system output. To fully comprehend the Laplace transform, some mathematical understanding of the concept is required. Its mathematical formulation is quite like the Fourier transform, but instead of $i\Omega$, a variable s is replaced which in general terms is a complex number. The transfer function can be written as:

$$H(s) = \frac{Y(s)}{X(s)}$$

Where $H(s)$, $X(s)$ and $Y(s)$ are the Laplace transforms of the impulse response, input and output of a system.

Similarly, for discrete time signals, the transfer function is expressed in terms of the Z-transform, which is also an extended concept compared to the Fourier transform, in which Z is generally a complex number, often expressed in "polar form" with a phase and magnitude (see Appendix A.2). Thus, for discrete time systems, the transfer function can be written as:

$$H(z) = \frac{Y(z)}{X(z)}$$

7.3 THE SYSTEM RESPONSE AND DIFFERENTIAL AND DIFFERENCE EQUATIONS

The previous section outlined the relationship between the frequency response and the transfer function of an LTI system. I refer to LTI systems in the rest of the text unless otherwise stated. The response of an important subclass of an LTI system can be described by a differential (for continuous time systems) and a difference equation (for discrete time systems). My aim is not to get caught up discussing differential equations and the difference equations. However, it is essential to realize that there is a straightforward relationship between such equations and the system transfer function. I assume that readers of this book have at least some knowledge of differentiation in mathematics (that is, calculating the function derivative). Most people learn the basics of calculus during their high school education. Here, I provide a brief reminder.

Let us assume $x(t)$ is a signal. As mentioned previously, in the context of mathematics, one can see $x(t)$ as a function of time. By differentiation in time, one means to calculate such a limit:

$$\frac{dx}{dt} = \dot{x}(t) = \lim_{\Delta t \to 0} \frac{x(t + \Delta t) - x(t)}{\Delta t}$$

Where $\frac{dx}{dt}$ and $\dot{x}(t)$ are commonly used to denote the first-order derivative. This indicates that for a small change in time, the signal amplitude changes. On can continue this procedure to obtain the second-order derivative (denoted $\dfrac{d^2x}{dt^2}$ and $\ddot{x}(t)$). This can continue for higher orders as well.

For the mentioned subclass of LTI systems, the system response can be described in terms of a weighted sum of the input and output and their derivatives up to some certain order.

For example, for a particular system, one could write:

$$\alpha_1 \dot{x}(t) + \alpha_0 x(t) = \beta_0 y(t) + \beta_1 \dot{y}(t) + \ddot{y}(t)$$

Where $x(t)$ and $y(t)$ are the input and output of the system, respectively. α_0, α_1, β_0 and β_1 are some constants. These constants are called the system parameters or coefficients. The maximum order of the derivative in the equation is two, and therefore the system is called a second-order system or the "system order" is two.

Differential equations are effective tools to describe the relationship between the input and output of a continuous time system. For a discrete time system, a difference equation is used to describe such a relationship. This means that the system response can be described in terms of a weighted sum of the input and output and their shifted versions in time.

Thus, such a relationship can be described as follows:

$$\alpha_1 x(n-1) + \alpha_0 x(n) = \beta_0 y(n) + \beta_1 y(n-1) + y(n-2)$$

Where $x(n)$ and $y(n)$ are the input and output of the system, respectively, and α_0, α_1, β_0 and β_1 are system parameters. $x(n-1)$ in this equation denotes a shifted version of $x(n)$ by one unit of time. Similarly, $x(n-2)$ denotes two steps of a shift in time. In this case, the system order refers to the maximum steps of the time shift in the difference equation.

7.4   TRANSFER FUNCTION AND DIFFERENTIAL AND DIFFERENCE EQUATIONS

Performing some of the mathematical operations on the signals in time can be easily represented in the frequency domain. For example, when we differentiate a signal in time, this corresponds to a multiplication of its Fourier transform by $i\Omega$ in the frequency domain.

This means that if the Fourier transform of $x(t)$ is $X(i\Omega)$:

$$x(t) \xrightarrow{\mathcal{F}} X(i\Omega)$$

the Fourier transform of $\dot{x}(t)$ will be (Oppenheim, Willsky, and Nawab, 1997) (page 328):

$$\dot{x}(t) \xrightarrow{\mathcal{F}} i\Omega.X(i\Omega)$$

The mathematical derivation of this formula is quite straightforward, but because my intention is to keep the math in this text to a minimum, I will not explain how to derive this relationship.

Now, referring back to the discussion on the Laplace transform in Section 7.2, we can write

$$\dot{x}(t) \xrightarrow{\text{Laplace}} s.X(s)$$

Similarly, one can generalize the concept and write

$$\ddot{x}(t) \xrightarrow{\text{Laplace}} s^2.X(s)$$

Now have another look at a hypothetical differential equation describing the relationship between input and output:

$$\alpha_1 \dot{x}(t) + \alpha_0 x(t) = \beta_0 y(t) + \beta_1 \dot{y}(t) + \ddot{y}(t)$$

Given the linearity of a Fourier transform (which essentially can be generalized to a Laplace transform) and what I outlined earlier, one can obtain the Fourier (Laplace) transform of both sides of the previous equation:

$$\alpha_1 s X(s) + \alpha_0 X(s) = \beta_0 Y(s) + \beta_1 s Y(s) + s^2 Y(s)$$

Now we can simplify this equation for the Laplace transform:

$$\frac{Y(s)}{X(s)} = \frac{\alpha_1 s + \alpha_0}{s^2 + \beta_1 s + \beta_0}$$

Or for the Fourier transform:

$$\frac{Y(i\Omega)}{X(i\Omega)} = \frac{\alpha_1 (i\Omega) + \alpha_0}{(i\Omega)^2 + \beta_1 (i\Omega) + \beta_0}$$

This is an interesting relationship, because $\frac{Y(s)}{X(s)}$ is actually the transfer function of the system. This means that for this class of LTI systems, the transfer function can be written as a ratio between two polynomials in s (Laplace transform) or $i\Omega$ (Fourier transform). One may ask what a polynomial is. In math, a polynomial is a function that can be written as a weighted sum of the function variable (in this case the s) and its integer exponents. For example, in the previous equation $s^2 + \beta_1 s + \beta_0$ is a polynomial in s (the function variable). Polynomials are quite convenient and exhibit nice mathematical features that facilitate some of the necessary operations.

One can also derive an interesting relationship between the difference equation and Fourier transform for discrete time systems. This means that if the Fourier transform of $x(n)$ is $X(e^{i\omega})$, the Fourier transform of $x(n-1)$ will be $e^{-i\omega} X(e^{i\omega})$. Similarly, the Fourier transform of $x(n-2)$ will be $e^{-2i\omega} X(e^{i\omega})$. Thus, if the hypothetical difference equation of the system can be written as:

$$\alpha_1 x(n-1) + \alpha_0 x(n) = \beta_0 y(n) + \beta_1 y(n-1) + y(n-2)$$

Due to the linearity of the Fourier transform (and the Z-transform) and what I outlined earlier, the transfer function can be written for the Z-transform as:

$$\frac{Y(z)}{X(z)} = \frac{\alpha_1 z^{-1} + \alpha_0}{z^{-2} + \beta_1 z^{-1} + \beta_0}$$

Or for the Fourier transform as:

$$\frac{Y(e^{i\omega})}{X(e^{i\omega})} = \frac{\alpha_1 (e^{i\omega})^{-1} + \alpha_0}{(e^{i\omega})^{-2} + \beta_1 (e^{i\omega})^{-1} + \beta_0}$$

As it is evident from this formulation, the transfer function in this case is also the ratio of two polynomials in z ($e^{i\omega}$ for the Fourier transform). Note that equation shows polynomials as functions of z^{-1} but this is not an essential discrepancy from what I stated earlier. If the numerator and the denominator of the fraction are multiplied by z^2, the polynomials will appear as a function of z.

7.5   THE SYSTEM RESPONSE IN THE FREQUENCY DOMAIN

Remember that the transfer function was the Fourier transfer of the impulse response in the frequency domain. Thus, it can be represented as a complex number with a magnitude and phase of the response. When one studies the properties of systems, the system response is often investigated in terms of its magnitude and phase response, which are actually the magnitude and response of the transfer function as a complex number in the frequency domain. This means that if one has the transfer function of a system, the transfer function can be calculated for each point along the frequency axis, and the magnitude and the phase of the transfer function would be functions of frequency. Figure 7.2 shows the magnitude and response of a system.

The magnitude response determines which frequency components will pass through the system and which components will be relatively attenuated. When the magnitude response is close to its maximum, the corresponding frequency at that point will pass through the system, and as it gets close to zero, the corresponding frequency will be attenuated. As can be seen in Figure 7.2, the frequency components in the range of 10 to 400 Hz will pass through the system, whereas those in the range above 400 Hz will be significantly attenuated. This means that if the system input has frequency components within the range of 10 to 400 Hz, the output will

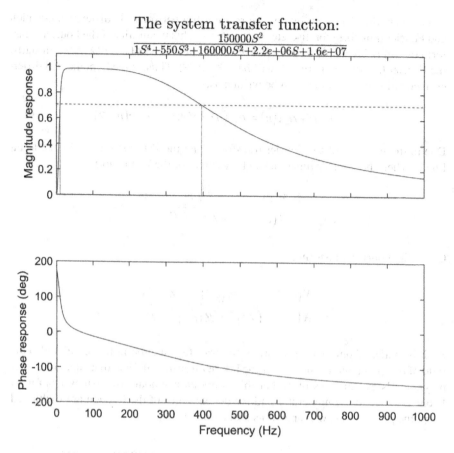

The system transfer function:
$$\frac{150000S^2}{1S^4+550S^3+160000S^2+2.2e+06S+1.6e+07}$$

FIGURE 7.2   The magnitude and phase response of a system

also have those components, whereas the energy of the frequency components in the range of above 400 Hz will be much lower in the output of the system. This is the basis of what we know as frequency-based filtering. If we know that our signal is within a certain frequency range, then the frequency components outside that range are the noise (which is undesirable) and we have to get rid of it. For example, the surface EMG is known to be within the range of 10 to 400 Hz; therefore, the frequency components out of this range should be filtered out. The half-power bandwidth of the system is determined by the frequency points at which the power of the output reaches half of the peak power of the system output at frequencies with the least level of attenuation (Papagiannopoulos and Hatzigeorgiou, 2011). These frequency points correspond to the points where the magnitude of the transfer function drops to $\frac{1}{\sqrt{2}}$ of the maximum of the transfer function. For example, as can be seen in Figure 7.2, the magnitude of the transfer function drops to $\frac{1}{\sqrt{2}}$ of its maximum at 10 and 400 Hz.

Very often, the magnitude of the transfer function is expressed in a logarithmic scale. This is the same scale I introduced in Chapter 1 as for decibels (db). Figure 7.3 shows the magnitude response of the same system in Figure 7.2, but this time, it is expressed in db. The half-power bandwidth of the system will then correspond to where the magnitude of the system drops by 3 db.

The phase response carries information about the time delay that a system will cause. Figure 7.4 shows the response of the system shown in Figure 7.2 to an arbitrary EMG signal. As is quite clear in Figure 7.4, the timeline of the system output is time-lagged with respect to the input.

In general, this time lag is a function of the frequency, but ideally, we would like the amount of the time delay to be the same across the entire frequency axis; otherwise, the system will distort the input signal in an undesirable way. Now the question is what the phase response should look like so that we can say that the amount of delay is constant across the entire frequency axis.

As was explained with Fourier transforms, a constant shift in time (time lag) will be transformed into a change in the phase, which can be shown to be a linear function of the frequency. Thus, we would like the phase of a system to be linear because

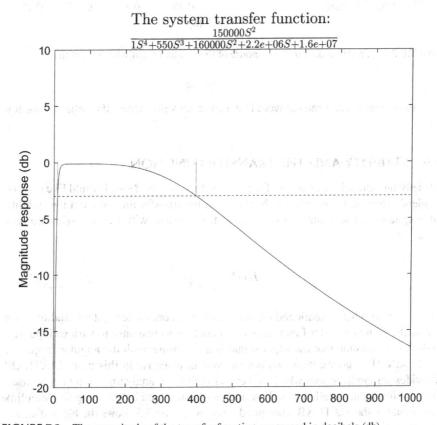

FIGURE 7.3 The magnitude of the transfer function expressed in decibels (db)

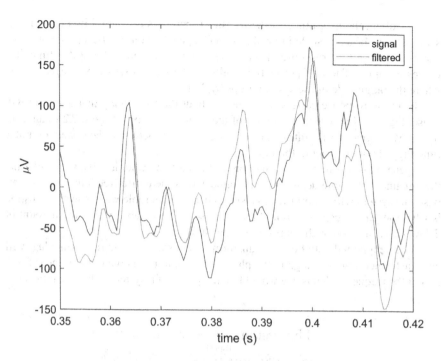

FIGURE 7.4　The time shift in the response of the system output with respect to the input

then the extent of the time lag would be a constant value across the entire frequency axis.

7.6　STABILITY AND THE TRANSFER FUNCTION

I briefly introduced the concept of stability in this chapter. Now, I would like to draw readers' attention to the relationship between the transfer function and the stability of a system. Let us assume a continuous time system with the following transfer function:

$$H(s) = \frac{1}{s^2 + 4}$$

According to what I mentioned earlier, this is a second-order system, and once we know the system transfer function, we can analyze its response to various input signals. Let us assume that the input signal is a sine wave with the angular frequency of 2 rad/s. How do we then find out the system response in this case? MATLAB[1] provides an environment which is very effective for analyzing system responses. The environment is called Simulink and can be accessed by invoking the Simulink command in the MATLAB[1] command window. Figure 7.5 shows the block diagram of the system implemented in Simulink.

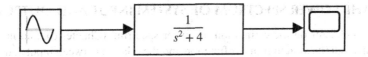

FIGURE 7.5 The schematic of simulating a system response in Simulink

After running Simulink, create a blank model, and from the library browser menu, drag the following items to the model page:

Transfer Fcn
Sine wave
Scope

Double-click in each block and set their parameters to replicate our hypothetical experiment. Note that the Transfer Fcn box accepts two polynomials as its input. The polynomials specify the coefficients of the polynomial in the numerator and denominator of the transfer function. For example, $s^2 + 4$ is specified by a polynomial in MATLAB[1] as an array of [1, 0, 4]. The numbers from the right side to the left side of this array represent the coefficients of s powered by zero to s powered by two. After setting up the model, you can run it to retrieve the response.

Interestingly, when one simulates the system response, the output unboundedly grows in time. This illustrates system instability, as the input to the system has been bounded (a sine wave with a limited range of [−1, 1]) and the system output becomes unbounded (increases without a limit).

One may wonder how I knew that if I simulate the system with a sine wave of 2 rad/s, the system becomes unstable. The answer comes from the form of a transfer function. The roots of the denominator in the transfer function (set the denominator to zero), $s = \pm 2i$, corresponds to a sine wave with 2 rad/s in the time domain. Another interesting observation is that if the input to the system had a different angular frequency other than 2 rad/s, the system response would have been bounded, showing a stable system. The roots of the transfer function denominator are called the system poles. The system poles can specify the inputs that can excite the system to instability. Conversely, the roots of the transfer function numerator are called the system zeros, and these can specify the input that can be hampered by the system. This particular angular frequency of 2 rad/s which resulted in an unstable behavior of the system is called the resonance frequency, and in this case, the system has resonated when triggered by the resonance frequency, but the system shows a stable response if triggered by any other frequency.

Along these same lines, system poles can be defined for discrete time systems. The difference, however, is that the transfer function for discrete time systems is presented by the Z-transform and stable poles lie within the "unit circle", meaning that the magnitude of the stable poles, which in general are complex numbers, must be smaller than 1. One may now appreciate why the angular frequency in the Fourier transform of discrete time systems is presented as $e^{i\omega}$. The frequency axis can be set around the unit circle, as it can show the periodic nature of the Fourier transform for discrete time systems.

7.7   THE POWER SPECTRUM OF SYSTEM INPUT AND OUTPUT

I wrote a chapter on the estimation of power spectrum (Chapter 6), and now after reading about the system transfer function and the relation between input and output, one may wonder whether there is any straightforward relationship between the system transfer function and the power spectrum of the input and output. This is a valid question, and luckily, with a rather straightforward algebraic procedure, one can show that the power spectrum density (PSD) of the input and output of an LTI system would be related as outlined here for continuous time systems (Papoulis, Pillai, and Unnikrishna, 2002) (pages 413 therein):

$$S_{yy}(i\Omega) = |H(i\Omega)|^2 S_{xx}(i\Omega)$$

Where $S_{yy}(i\Omega)$ and $S_{xx}(i\Omega)$ are the PSD of the system input (x) and the output (y) and $|H(i\Omega)|$ is the magnitude of the system frequency response.

Similarly, for discrete time systems, one can write (Papoulis, Pillai, and Unnikrishna, 2002) (page 424):

$$S_{yy}(e^{i\omega}) = |H(e^{i\omega})|^2 S_{xx}(e^{i\omega})$$

The bottom line, and a point to remember, is that the power of the output in the frequency domain is proportional to the square of the magnitude of system frequency response multiplied by the input power spectrum.

7.8   THE TRANSFER FUNCTION AND PARAMETRIC ESTIMATION OF THE POWER SPECTRUM

For discrete time signals, we have a practical approach that can provide a nice estimate of the signal PSD based on parametric methods. In Section 6.3.1, I mentioned that the length of a signal limits the frequency resolution of the PSD estimate. I also briefly mentioned the issues regarding estimation variance and how to reduce it at the cost of losing the frequency resolution. I mentioned that the transfer function is closely tied to the frequency response of a system. Now imagine modeling the signal as if it is the result of a system response to a noisy input known as white noise. If the structure of the system transfer function is known, the system transfer function can be estimated based on the modeled signal, and the PSD of the signal can readily be derived from the system transfer function.

Now one may ask what white noise is. White noise is a hypothetical random noise that covers the entire range of the frequency spectrum and theoretically has a constant power across the entire spectrum. Thus, the PSD of white noise is a constant value (its variance) across the entire spectrum, and it is neutral with respect to any part of the spectrum. Figure 7.6 depicts this hypothetical scenario, where the signal is the output of a system with a known transfer function and white noise as the input.

If one can implement a procedure as such, the frequency resolution is not a problem anymore, because the PSD of the signal can be fully described by the

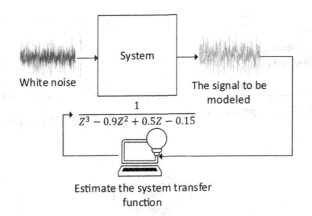

FIGURE 7.6 The signal modeled as the output of a system with a known structure of the transfer function and white noise as the input

frequency response of the system. Now the question is how to figure out what the transfer function should be. If one sets some reasonable constraints on the structure of the transfer function, the signal itself can be used to determine the system transfer function coefficients (parameters). In other words, one models the signal by deriving the system parameters. That is why these methods are called "parametric". If one assumes that the transfer function belongs to an LTI and causal system, three structural forms can be envisioned. The system may have only zeros without any pole. In this case, the system will be a moving average (MA) system. This means that the output of a system at each point in time is determined by a weighted average of the current and past samples of the input. Other possible structures would be a system that has only poles and no zero. In this case, the system response at each time point is regressed in the past sample of the output and the current input, and that is why this system would be called an autoregressive (AR) system. The other possibility would be a combination of the previously mentioned models, and it is called an autoregressive moving average (ARMA) system. This system has zeros and poles, and the output can be derived as a weighted average of the current and past inputs, as well as previous samples of the output. Figure 7.7 illustrates these possibilities.

Now the question is how to decide which of these possibilities should be adopted and the model order. This question is tenable in any modeling scenario. How do we know whether we have a good model to describe the phenomenon under investigation? Generally, a good model is a model that has the smallest number of parameters while it can fit the signal to a greater extent. Some indices of the goodness of fit consider these two factors (parameter number and fit criterion). For example, "adjusted coefficient of determination" and "Akaike information criterion" are among such indices. A data analyst will play with the model order and model structure and find the one that is optimal in terms of parameter number and fit criterion. The validity of the parametric method depends on the appropriateness of the model to describe the signal. No good model means no accurate estimate of the PSD.

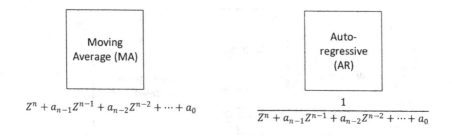

FIGURE 7.7   The general form of the system transfer function of a moving average system, an autoregressive system and an autoregressive moving average system

This topic will not be discussed further than this. Understanding how to estimate the model parameters requires quite the algebraic workout, and my intention is to keep this book free from math as much as possible.

NOTE

1. These MATLAB files are available at https://www.crcpress.com/9780367207557.

REFERENCES

K. Y. Billah and R. H. Scanlan, "Resonance, Tacoma Narrows bridge failure, and undergraduate physics textbooks," *Am. J. Phys.*, vol. 59, no. 2, p. 118, 1991.

D. Farina, F. Negro, and N. Jiang, "Identification of common synaptic inputs to motor neurons from the rectified electromyogram," *J. Physiol.*, vol. 591, no. 10, pp. 2403–2418, 2013.

A. V Oppenheim, A. S. Willsky, and S. H. Nawab, *Signals and Systems* (2nd ed.). NJ: Prentice Hall, 1997.

G. A. Papagiannopoulos and G. D. Hatzigeorgiou, "On the use of the half-power bandwidth method to estimate damping in building structures," *Soil Dyn. Earthq. Eng.*, vol. 31, no. 7, pp. 1075–1079, 2011.

A. Papoulis, S. U. Pillai, and S. Unnikrishna, "Probability, Random Variables and Stochastic Processes," *Technometrics* (4th ed.). McGraw-Hill, 2002.

8 Filters

Now that I have introduced systems and their transfer function, we can turn to a category of systems known as filters—in particular, frequency-based filters. Generally, filtering is the process of removing noise from the observed signal and retrieving the desired signal. If one can assume a model underlying the desired signal, some computational algorithm such as Wiener and Kalman filters are used to remove, or at least reduce, the effect of noise and retrieve the desired signal. This chapter will discuss frequency-based filters—when we have some idea about the frequency representation of our desired signal and we know that the power of the observed signal outside of a specific frequency range is due to noise, the aim is to attenuate the signal power in that range.

8.1 IDEAL FREQUENCY RESPONSE OF FILTERS

As mentioned earlier, when one has an idea of the spectrum of the desired signal and knows that the desired signal does not have a significant power in a certain frequency band, reducing the power of the observed signal in that specific range will remove or attenuate the noise. Now one can imagine several scenarios: i) the signal is expected not to have significant power above a certain frequency threshold; ii) the signal is expected not to have significant power below a certain frequency threshold; iii) the signal power is mainly limited to a certain frequency band; and iv) the noise is mainly condensed into a certain frequency band. In the first scenario, the filter should allow frequency components lower than a given limit to pass without significant attenuation, whereas the frequency components above that limit should be attenuated heavily. Therefore, the filter is known as a low-pass filter, and the frequency limit of pass/no-pass is known as the cut-off frequency of the filter. Conversely, in the second scenario, the filter should allow frequency components higher than a limit to pass without significant attenuation, whereas the frequency components below that limit should be attenuated heavily. Therefore, the filter is called a high-pass filter. The third scenario is a combination of the first and second scenarios, where the signal is expected to be within a certain band and any frequency component outside that band should be attenuated heavily. Therefore, the filter is known as a band-pass filter, and the band is within a low and high cut-off frequency. Finally, the last scenario is the opposite of the third one and hence, the filter is known as a band-stop filter. If the stopband is very narrow, the filter is known as a notch filter. Figure 8.1 illustrates how the frequency response of filters should ideally look.

What in practice one can achieve with frequency-based filters does not have a sharp and clear-cut frequency response as shown in Figure 8.1. There will always be a transition phase around the cut-off frequency. The cut-off frequency is often defined as a threshold where the power of the output signal drops to half of the power of components with no attenuation (Widmann, Schröger, and Maess, 2015) (often

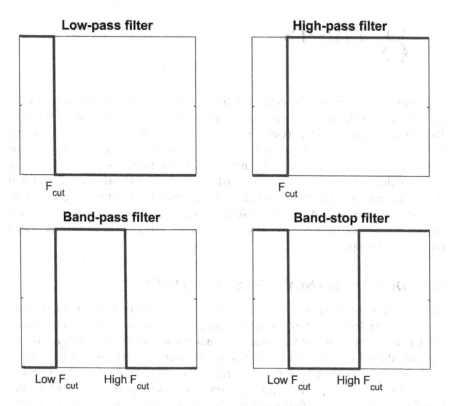

FIGURE 8.1 The ideal magnitude of the frequency response of the frequency-based filters: a low-pass filter, a high-pass filter, a band-pass filter and a band-stop filter.

where the magnitude of the frequency response is 1, the filter does not attenuate the frequency components in that range). I already stated that the power spectrum of the output is proportional to the square of the magnitude of the system frequency response, and therefore, the cut-off frequency is the frequency limit where the magnitude of the frequency response drops to $\frac{1}{\sqrt{2}}$ of its magnitude when no attenuation is applied (often where the magnitude is 1).

The frequency response of a filter in practice can be divided into three segments, namely, a passband, a transition band and a stopband. The passband determines the interval across the frequency axis, where the filter allows the input signal within that range of the frequency passing through the filter. The magnitude of the frequency response is close to 1, considering a permissible range of oscillation around 1, which is known as the "ripples" in the passband. Conversely, the stopband is the opposite and determines the interval across the frequency axis where the filter does allow the input signal within that frequency range passing through the filter. The magnitude of the frequency response is close to zero, considering a permissible range of oscillation around zero, which is known as the ripples in the stopband. The transition band, as

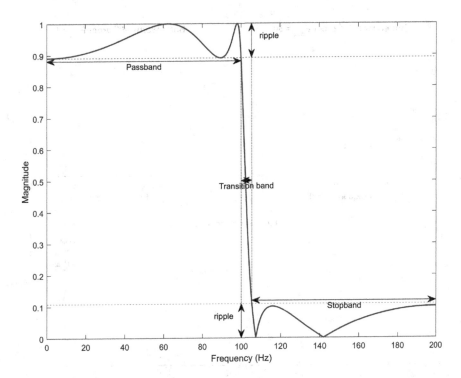

FIGURE 8.2 A typical practical frequency response of a low-pass filter with a passband, a transition band and a stopband; the figure also shows ripples in the passband and stopband

the name indicates, refers to the interval between the passband and the stopband. As opposed to the ideal response of a filter, the transition band cannot be squeezed into one single frequency point, and the difference between the edges of the passband and stopband is non-zero. The filter attenuation in the transition phase is between its attenuation in the passband and stopband, but it is still higher than the ripples in the passband and lower than the ripples in the stopband. Figure 8.2 shows a typical response of a low-pass filter in practice. The response of high-pass, band-pass and band-stop filters can be seen in Figure 8.2.

8.2 FILTER ORDER

I have already defined the system order in Chapter 7. The filter order refers exactly to the same concept. As the filter order increases, the transition band of the filter becomes sharper and sharper, meaning that the edges of the passband and stopband get closer to each other. At first glance, one may conclude that a higher filter order is necessarily a better feature of a filter, because the practical response of the filter tends to approach its ideal response. However, a filter with a higher order requires more computational resources for its implementation and may even lead to computational instabilities in rendering the response. If the structure (I will outline what

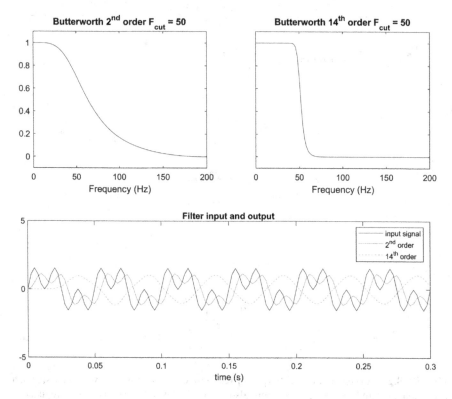

FIGURE 8.3 The frequency response of two filters with different orders and their time response for a synthetic signal

I mean by this in the following section) of two filters is similar and the only difference between them is their order, a sharper response in the frequency domain is equivalent to a wider effect in time, and likely more distortion in time.

Figure 8.3 shows the frequency response of two low-pass filters with the same cut-off frequency (50 Hz) but with a different order (2 versus 14). The input to these low-pass filters is a synthetic signal with two frequency components at 20 and 60 Hz. As can be seen, the higher-ordered filter attenuated the 60-Hz component quite a lot and the output looks like a sine wave with 20 Hz, while the two frequency components can still be seen with second-order filter. However, the extent of the time delay of the filter output with respect to its input is longer with the fourteenth-order filter.

8.3   FILTER DESIGN

Knowing the filter cut-off frequency and filter order does not fully specify a filter. Furthermore, the cut-off frequency is not defined for a specific structure of filters (e.g., Chebyshev type I). For example, Figure 8.4 shows the magnitude of the frequency response of two low-pass filters with the same passband edge frequency and the same order. As can be seen, the filter responses are quite different.

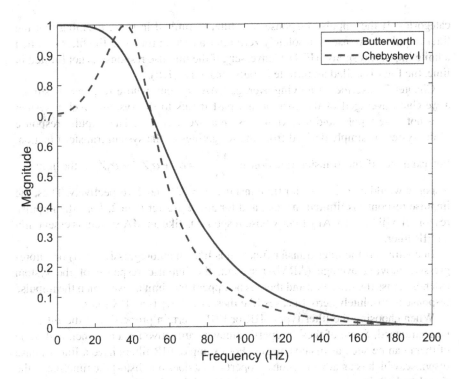

FIGURE 8.4 The magnitude of the frequency response of two filters with the same pass-band edge frequency and filter order

To fully specify the filter response, one can impose other constraints such that only a unique filter can meet all the applied constraints. For example, one might want a filter whose magnitude of the frequency response is maximally flat and does not have any bumpiness (i.e., ripple) in the response, or one may want a filter response that has the shortest transition band for a given order, or may want a filter with a maximally linear phase response.

The process of addressing these extra constraints is known as filter design. There are already some very well-established filter design techniques whose properties have been well-studied, but a keen designer can still customize an innovative filter design with specific properties that suits a particular application. The filter design actually gives the coefficients of the filter transfer function. Once one has the filter transfer function, the filter response to a specific input signal can be calculated.

Similar to what I mentioned regarding systems, filters can also be continuous- or discrete time systems, but here I limit my discussion to discrete time filters because they are relevant to digital signal processing, meaning that when processing signals on a computer, we deal with discrete time systems. Discrete time filters are also known as digital filters. As I explained earlier, the response of discrete time filters can be written as a difference equation, as outlined in Section 7.4. I mentioned that the transfer function is very closely tied to the frequency transformation of the system impulse response. In light of this notion, the filters can be divided into two categories. If the impulse response of a filter is limited in time, meaning that the filter impulse response is absolutely zero after a certain time lag, the filter is called a finite impulse response (FIR). Conversely, if the impulse response is not limited in time, the filter is called an infinite impulse response (IIR).

Chapter 7 described a moving average (MA) system, whose response is simply a weighted averaged of the current and past inputs to the system. Such a system does not have a pole, and it is known as an all-zero system. The impulse response of the system is simply derived from the weightings of the system transfer function. For example, if the transfer function is $\dfrac{Y(Z)}{X(Z)} = \alpha_0 + \alpha_1 Z^{-1} + \alpha_2 Z^{-2}$, the impulse response would be α_0, α_1, α_2 for time instance $n = 0$, 1 and 2, respectively. Thus, the impulse response is limited in time, and for any n greater than 2, $h(n)$ (the impulse response) will be zero. Any filter whose response is like an MA system is essentially an FIR filter.

In contrast, if the filter transfer function is like an autoregressive (AR) or autoregressive moving average (ARMA) system, the impulse response of the system extends across the time axis and there is no upper time limit above which the impulse response is absolutely zero. In cases like these, the filter is an IIR filter.

When choosing the filter type—IIR or FIR—certain properties of the filter can be guaranteed. Some of these properties may be attractive for a designer, and some of them can be seen as disadvantages. For example, FIR filters have a linear phase response, which is an advantageous property, as it does not distort the timeline of the signal and all frequency components would be delayed in time to the same extent. Another important property of the FIR filters is guaranteed stability. As explained earlier, the FIR filters are all-zero systems and have no pole; therefore, the filer is always stable even if the input is subjected to some distorting noise or there is some computational disturbance (e.g., quantization error) in the calculation of the filter response. On the other hand, in order to get a very short transition band of the filter, the transfer function of the filters will need to consist of many more coefficients than an IIR system with a similar transition band. Thus, one needs to consume more computational resources to calculate the filter response compared to an IIR filter with similar frequency response.

The IIR filters can be quite efficient in saving computational resources when the output of the filter is calculated, but a linear phase cannot be guaranteed across the entire frequency axis, and they are susceptible to computational disturbances, which may result in instability of the filter response. The designer then has to decide between inherent stability and linear phase against computational efficiency.

Depending on the choice of the filter response, IIR or FIR, there are specific algorithms to design a filter with specific properties. Here, I do not intend to introduce various algorithms for designing filters but will briefly mention a few of them and some of their known properties. A common method for designing FIR filters is known as the window method (Kiraç and Vaidyanathan, 1998). One starts with a desired frequency response and then calculates the inverse Fourier transform for a given response to obtain the desired impulse response. The desired response is not necessarily limited in time, but in certain conditions if the impulse response is decaying in time, one may truncate the response to limit it within a certain time threshold. The frequency response of a filter designed with this technique will not be identical to the desired frequency response, but it may satisfy the main requirements posed by the designer.

Technically, instead of considering a specific filter order, a designer can start by characterizing the desired filter response and imposing a "minimum order constraint". This means that once the designer knows the passband frequency, stopband frequency and admissible range of variation of the filter attenuation in the passband and stopband, the designer can impose a constraint such that the designed filter has a minimum order for a given filter structure.

There are some well-established algorithms for filter design, and each imposes specific constraints to the filter response. For example, if one imposes the maximal flatness to the magnitude of the filter frequency response, the resulting filter is called a Butterworth filter. The magnitude of the Butterworth frequency response is very flat, and no ripple can be seen either in the passband or the stopband of the filter (see Figure 8.4). Alternatively, if one wants to obtain a filter with the shortest transition band for a given filter order while the distortion to the linearity of the phase response is minimal, the resulting filter would be called a Chebyshev filter. However, this type of filter will have some ripples either in the passband (Type I) (see Figure 8.4) or in the stopband (Type II).

So far, I have focused on the design of low-pass filters, so one may wonder about the high-pass, band-pass and band-stop filters. How does one design those filters? Technically, once one is able to design a low-pass filter with a given set of properties, designing the other ranges of frequency selection (i.e., high-pass, band-pass and band-stop) with similar properties is just a matter of transforming the filter transfer function. I do not delve into this topic because the mathematical details of the procedure are beyond the scope of this section, but those who are interested to know more can refer to (Oppenheim, Schafer, and Buck, 1989) (page 430 therein).

8.4 FILTER PHASE RESPONSE

The system phase response was discussed in Chapter 7. The content there is also applicable to filters. As can be seen in Figure 8.3, the filter output is delayed with respect to its input. I have already shown that this time delay is related to the phase response of the system (in this case, the filter). Is there a way to get rid of this time delay? Well, the answer is yes and no. It just depends on our limitations in working with causal systems. Imagine a system that works in real time. A real-time system constantly receives the input, and the output of the system should be calculated as quickly as possible. In this case, the output of the system at a time point can only be a function of the current and past inputs, as well as past outputs. In a scenario like this, the time delay imposed by the filter cannot be avoided. The best thing we can aim for is to design a system (filter) that has a linear phase (at least in the passband range) such that the system does not cause any distortion to the signal timeline.

However, sometimes we do some offline analysis, meaning that we have collected a dataset and we analyze it later. In such a scenario, there is no limitation on the time of the output calculation, and at each time point, we can calculate the system output, not only based on the past and current inputs but also input samples recorded after a specific time point. In other words, we can access future inputs. A system that accesses future inputs is a non-causal system and can only be realized if one is working with a set of data that has been already collected.

In filtering, bidirectional or zero-phase filtering allows for nullifying the phase response as if the filter response does not induce a delay in the output. Bidirectional filtering is based on a very nice property of the transfer function in which the sequence of the coefficients in the transfer function is reversed, and one obtains a filter that has an identical magnitude response, but the phase response is a mirror image of the phase response of the main filter. If the two filters are applied sequentially, the overall system has a phase of zero and no time delay is applied. Figure 8.5 illustrates this concept.

In MATLAB[1], the "filtfilt" function applies bidirectional filtering. One should note that when filters are applied, they have a transient response at the edge of the signals before their response settles to what one would expect. The same limitation is applicable to bidirectional filtering (Gustafsson, 1996). The other important note about bidirectional filtering is that with this technique we actually apply two filters; therefore, the overall filter order is two times that of the main filter.

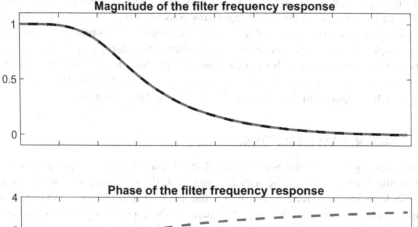

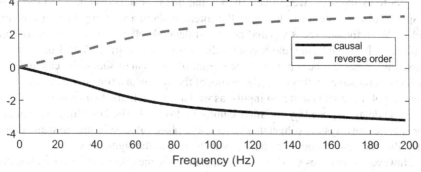

FIGURE 8.5 Magnitude and phase response of a causal filter and its reversed version; note the identical magnitude response and mirror phase response in the reversed system

8.5   IMPLEMENTATION OF FILTERS

These days, the rapid development of electronic systems allows the production of small and powerful processors that can be embedded in various devices and enable rather heavy computation. These hardware units are known as embedded systems. These devices perform computational tasks such as analog to digital conversion and digital filtering on a small-scale device.

Even though the processors are becoming more powerful in performing computational operations, they are typically weaker than the processors running in a brand-new computer. Additionally, these devices run on batteries, and heavy computation loads will consume higher levels of energy sources. Therefore, it is crucial that the computational tasks that are deployed to these devices work optimally with minimal memory usage. For example, if one can reduce the number of arithmetic operations (adding, multiplication) in the implementation of a filter and, as a result, the computational power needed, the device can save more energy and can work faster. Similarly, it is also beneficial if one uses fewer memory units in the implementation of embedded systems.

Here is an example to illustrate the concept. I previously mentioned that calculating a digital filter output involves solving a difference equation for a given input. Let us assume that the difference equation can be written as follows:

$$y(n) = \underbrace{\alpha_0 x(n) + \alpha_1 x(n-1)}_{input\ terms} - \underbrace{\left(\beta_1 y(n-1) + \beta_2 y(n-2)\right)}_{output\ terms}$$

This is a second-order filter, and the difference equation shows that at each point in time, the output of the filter equals a weighted sum of the input at the time and its past sample as well as the last two samples of the output itself. Now let us draw a block diagram (Figure 8.6) that shows the required steps to calculate the output of this difference equation.

What Figure 8.6 shows is that for a given input $x(n)$, the processor must multiply the input at that time by α_0; the processor must also store the previous sample of input $x(n-1)$ and multiply that by α_1. This would give the input terms in the equation. Additionally, the processor must have stored the last two samples of the output $y(n-1)$ and $y(n-2)$ and multiplied them by β_1 and β_2, respectively. The summation of these would give the output terms in the equation, and now if one subtracts the output terms from the input terms, the output for the current time point is readily available. To do this, one would need to have three units of memory to store the previous sample of the input, as well as the last two samples of the output itself.

If one looks at Figure 8.6 closely, one can see that the entire system can be divided into two subsystems: one that works with the input, let us call it subsystem I, and the other one that works with the output, let us call it subsystem O. One can actually derive a difference equation for the subsystems I and O. Actually, they themselves can be seen as a system.

$$y(n) = \underbrace{\alpha_0 x(n) + \alpha_1 x(n-1)}_{subsystem\ I}$$

$$y(n) = \underbrace{x(n) - \left(\beta_1 y(n-1) + \beta_2 y(n-2)\right)}_{subsystem\ O}$$

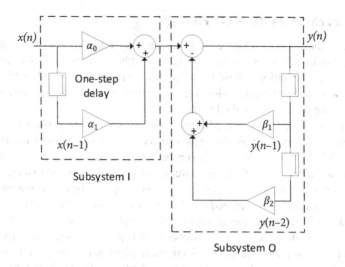

FIGURE 8.6 A block diagram showing the required steps to find the filter output. Modified from Oppenheim, Willsky, and Nawab, 1997 and Oppenheim, Schafer, and Buck, 1989

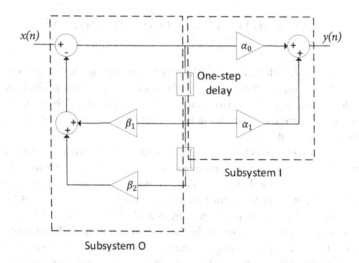

FIGURE 8.7 Changing the sequence of the subsystems in a cascade. Because the two subsystems are linear, changing the sequence does not change the relationship between the input and output of the entire system. Modified from Oppenheim, Willsky, and Nawab, 1997 and Oppenheim, Schafer, and Buck, 1989

It is straightforward to verify that both subsystems I and O are linear systems, and therefore if we change their order in a sequence (cascade) of systems, the relation between the input and output of the entire system will be preserved. This means that if the input is first applied to the subsystem I and then the output of the I is applied to the O, the resulting output is the same if we do it in a reverse order, with O first and I second. As can be seen in Figure 8.7, the number of memory units in this case is only two. Now imagine that if the filter order were higher, one would have saved even more memory units.

This is a typical example that can be found in many textbooks (Oppenheim, Schafer, and Buck, 1989). And with a little trickery, one can optimize how digital systems (e.g., a filter) can be implemented on the hardware and save some resources. Figure 8.6 shows an implementation that is known as "direct form I", and Figure 8.7 shows "direct form II". This topic is quite more extensive than what I outlined earlier, but this served as a useful overview.

8.6   SPATIAL FILTERS

Mathematically speaking, signals are functions of time, and for each point in time, we could find a corresponding value of the signal amplitude. In addition, signals in this case are one dimensional, meaning that for each point in time, there is only one single scalar value for the corresponding signal amplitude. However, signals in general are not limited by these constraints. Signals can, for example, be multidimensional, meaning that for each point in time, the signal is not represented by a single scalar amplitude but is represented as a vector with more than one element. Signals in general do not need to be a function of time either. For example, a signal can be a function of spatial coordinates.

In eyes of a data analyst, an image is a two-dimensional "spatial signal" that varies across the pixels of an image. If an image is a gray-scale image, the amplitude of the signal will be the brightness of the pixel, whether that pixel is black, white or something in between. If an image is colored, then the signal should be represented as a three-dimensional vector indicating the intensity of red, green and blue in each pixel of the image. Thus, the basic concepts of signal processing introduced in previous chapters can be generalized for signals that are a function of more than one independent variable. For example, images are functions of the x and y coordinates of a pixel in the image. Furthermore, the Fourier transform can be generalized for multidimensional signals. Instead of presenting the sampling frequency in terms of samples per second, in an image, one would state "dots per inch" (dpi)—often known as the image resolution. Similarly, one can introduce the spatial frequency, representing the changes that a spatial signal has across the coordinates of the image pixels.

In recordings of biological signals, one can record over a matrix of electrodes and thus obtain the spatial pattern across the grid of electrodes—for example, in electromyography (EMG) recording (Gallina, Merletti, and Gazzoni, 2013). In such cases, one obtains the spatial pattern of a signal that also varies in time, and therefore it will be a spatiotemporal signal. Figure 8.8 shows the spatiotemporal pattern of activity over the trapezius, which can be recorded with a grid of electrodes covering the surface of the muscle.

Filters can also be generalized to be applied to spatial signals, and they can limit the extent of spatial frequency in the filtered image. Similar to frequency-based filters, the spatial filters can also be low-, high- or band-pass filters, but instead of filtering the temporal frequency, they would filter the spatial frequency. Figure 8.9 shows a synthetic image that has two frequency components across the

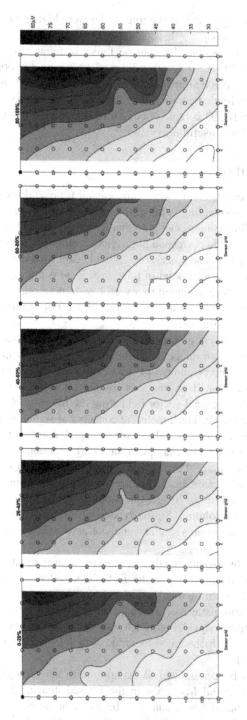

FIGURE 8.8   Spatiotemporal pattern of monopolar electromyographic activity of the trapezius muscle during a repetitive dynamic task

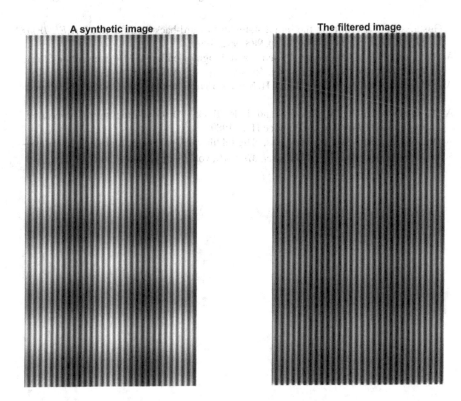

FIGURE 8.9   A synthetic image and its spatially high-pass filtered version

X-axis (i.e., 5 and 60 Hz) and one frequency component across the Y-axis (i.e., 5 Hz). The image is spatially high-pass filtered with a cut-off frequency of 30 Hz across the X- and Y-axes. As can be seen in Figure 8.9, low-frequency components across the X- and Y-axes are very clear, meaning that if one looks at the horizontal and vertical directions of the image, dark and bright bands across the X, Y axis is quite clear. However, this pattern is quite attenuated in the filtered version of the image. Please note I do not mean the narrow bright and dark bands but the wider bands which are visible in the original image but not in the filtered image.

An in-depth look at spatial signals and spatial filters is beyond the scope of this book. Here I just wanted to draw readers' attention to the fact that signals can cover a much wider field than a single-value observation across time.

NOTE

1. These MATLAB files are available at https://www.crcpress.com/9780367207557.

REFERENCES

A. Gallina, R. Merletti, and M. Gazzoni, "Uneven spatial distribution of surface EMG: What does it mean?" *Eur. J. Appl. Physiol.*, vol. 113, pp. 887–894, 2013.

F. Gustafsson, "Determining the initial states in forward-backward filtering," *IEEE Trans. Signal Process.*, vol. 44, no. 4, pp. 988–992, 1996.

A. Kiraç and P. P. Vaidyanathan, "Theory and design of optimum FIR compaction filters," *IEEE Trans. Signal Process.*, vol. 46, no. 4, pp. 903–919, 1998.

A. V Oppenheim, A. S. Willsky, and S. H. Nawab, *Signals and Systems* (2nd ed.). NJ: Prentice Hall, 1997.

A. V. Oppenheim, R. W. Schafer, and J. R. Buck, "Discrete-time signal processing," *Englewood Cliffs* (vol. 2). Prentice Hall, 1989.

A. Widmann, E. Schröger, and B. Maess, "Digital filter design for electrophysiological data— A practical approach," *J. Neurosci. Methods*, vol. 250, pp. 34–46, 2015.

Appendix A.1: A Brief Introduction to MATLAB

MATLAB[1] is a computational software package used by a wide range of scientists and engineers. In no way can this short introduction be a comprehensive overview of MATLAB functionalities, but it could be a starter for someone not familiar with it. MATLAB documentations are very extensive and the best source of information for learning how to work with it. Additionally, MathWorks, the producer of MATLAB, has provided a framework for MATLAB users to share code and consult with more experienced users.

The MATLAB documentation states: "MATLAB® is the high-level language and interactive environment used by millions of engineers and scientists worldwide. The matrix-based language is a natural way to express computational mathematics". So first and foremost, MATLAB is a high-level programming language that simplifies the process of programing quite a lot. Additionally, MATLAB has special capabilities to run matrix-based calculations; therefore, MATLAB programmers try to reform any computational problem into a matrix-form problem to benefit from MATLAB's power point.

When running MATLAB, you will see a window composed of a few subwindows. Figure A.1.1 shows frequently used subwindows. The most important of all is the command window, where one can invoke MATLAB commands. The command window will display a prompt marker (>>) when it is ready to receive a new command. At its simplest, MATLAB can be used as a calculator. If any computational statement is inserted into the command window, MATLAB will perform the calculation and display the results.

As with any other programming language, one can set some variables and work with them. MATLAB can work with various types of variables, such as numerical variables, which could simply be a scalar number or a multidimensional matrix. The following lines show how to set a variable and assign a specific value to it. If one types the following command in the command window and presses the Enter key, MATLAB starts working on the command.

```
>>a=2;
```

This statement assigns a value of 2 to a variable called "a", and the semicolon at the end of the command stops MATLAB's verbosity; otherwise, MATLAB would have echoed the value of the "a" like so:

```
a=2
```

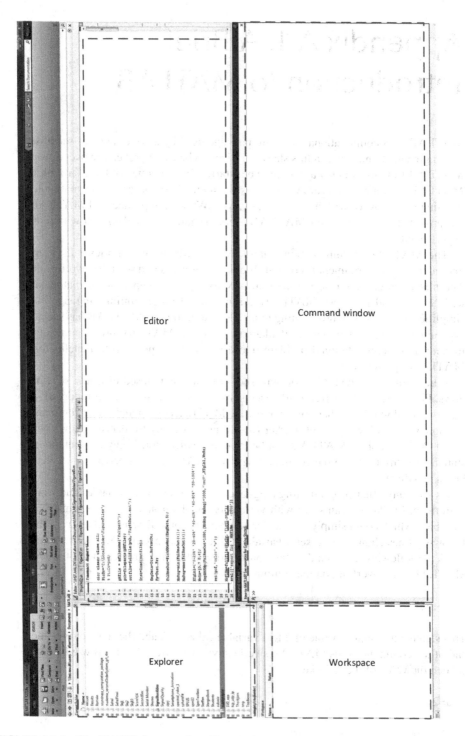

FIGURE A.1.1 The MATLB framework and its subwindows: command window, explorer window, workspace and editor

One can also set a vector or matrix in MATLAB, called a MATLAB array:

```
>>a=[1 2 4];
```

By running this command, "a" will be assigned a row vector with three elements. "a" is a row vector, which means that it has one row and its number of columns equals its number of elements. One can also assign a column vector to "a" such that

```
>>a=[1; 2; 4];
```

Alternatively, one can use commands and operators, which are often used to implement linear algebraic calculations. If one types the following command:

```
>>b=a';
```

The operator ' will transpose (swap the rows and columns of a matrix) of the "a" and then assign the resulting vector to "b". In this case, if "a" is a row vector, "b" will be a column vector and vice versa.

Now imagine you want to create a matrix with a few rows and columns in the MATLAB environment.

```
>>c=[1 2 3; 5 4 1];
```

In this case, "c" has two rows and three columns. The first row will contain [1 2 3] and the second row will be [5 4 1].

Now imagine you want to know what the element of "c" is, as well as the second row and column:

```
>>c(2,2)
ans=
4
```

Note that even though this seems to be useless, when one sees the entire matrix and can already see the element on the second row and column is 4, if the size of the matrix is much larger and seeing the entire matrix at a glance is not possible, this functionality is quite useful. Additionally, if one needs to do any programming and change this single element, this is the way to go.

In this case, "c" is two-dimensional with rows and columns, but one can also imagine higher-dimensional matrices; for example,

```
>>d=rand(2,3,5);
```

In this example, "d" will be a randomly generated three-dimensional matrix. The "d" has two, three and five sets of elements along the first, second and third dimensions, respectively.

If the command window contains a lot of commands, as ours currently does, you can clear the command window as follows:

>> clc

This will make the command window appear blank, but the variables that one assigned to it will be kept in the computer's memory. If one looks at the workspace window, a list of assigned variables can be seen.

Now let us do something slightly more complicated. Imagine that you have a list of students' marks in a few courses and you want to calculate the average mark of each student, but the weighting of each course is different. In this case, let us assume we have three students and two courses. The first course counts two times as the second course. How do we calculate the average? The mark of the first course should be multiplied by 2 (weighting two times more) and added with the mark of the first course, and the result of the summation should now be divided by 3. In other words, the mark of the first course should be multiplied by 2/3 and the second one by 1/3 and then they should be summed up.

This can be done in a number of ways in MATLAB—you could do all the calculations for each student separately, but that is not optimal, and if the number of students and the courses are quite large, this process is difficult and prone to error. How else can we do this simple calculation? A good way of doing it in MATLAB is through matrix multiplication. Here, imagine that the marks are between 0 and 20, where 0 is the absolute minimum and 20 is the highest mark.

```
>>CourseWeighting=[2/3; 1/3];
>>StudentsMark=[13 15; 12 17; 16 14];
>>Average= StudentsMark* CourseWeighting
Average =
13.6667
13.6667
15.3333
```

As you can see, this way of doing the calculation is much more effective, especially if the numbers of students and courses are large. Note that in this calculation, the operator * is used to perform a matrix multiplication. Note that the matrix multiplication imposes some constraints on the matrix size such that the number of columns in the first operand ("StudentsMark") should equal the number of rows of the second operand ("StudentsMark"). If this condition does not hold, MATLAB will issue an error message; for example,

```
>>CourseWeighting=[2/3; 1/3; 1/2];
>>Average= StudentsMark* CourseWeighting
```
*"Error using ***
Incorrect dimensions for matrix multiplication. Check that the number of col-
umns in the first matrix matches the number of rows in the second matrix.
To perform elementwise multiplication, use '.'."*

Thus, when one works with MATLAB commands, the theoretical limitations of what one does should be borne in mind.

As you can see, the code snippet composed of three lines indicates that they are run one after another. Can we achieve the same thing with just a single command? Here is where the MATLAB editor comes handy, and you can create something that MATLAB knows as an m-file.

If you type the following command in the command window, the editor will open. If from the file menu, a new file is selected, the same thing happens.

```
>>edit
```

Now we can write the earlier commands in the editor and save it as an m-file in the current directory. Name the m-file "CalcStudentAverage" and save it in the MATLAB current directory; typing the file name in the command window will run the commands in a batch and the results will be shown like before.

```
>> CalcStudentAverage
Average =
13.6667
13.6667
15.3333
```

CalcStudentAverage is called a script, and any variable that is set in the script can also be seen in the MATLAB workspace. Now imagine that you have several lists of students and courses and you need to run the same procedure for all of them. In this case, it is not optimal to create a script for each student. Or imagine that you have many such scripts—it could be the case that two variables in two different scripts are accidentally given identical names, even though the two variables were supposed to keep two different entities. In this case, the first script to run may change the value of that variable and the second script works with a variable that does not contain the correct values. Thus, it is not a good idea to keep the same workspace for all of the scripts. What can we do then?

We can save the m-file as a MATLAB function. Functions may need some input argument(s) and perhaps some output argument(s). The workspace of functions is independent from the command window workspace, and any variable assigned in the function can only be seen by that function, unless it is globally defined (global variables are beyond the scope of this appendix). How does one create a function in the editor then?

The template (syntax) is as follows:

```
function [output arguments]=function_name(input arguments)
```

Now imagine we want to run a procedure to calculate the average—the function will need to have access to the students' marks and the course weighting to calculate this. All the information that the function needs to work with should be passed on to the function as input argument(s), and the results of the calculation should be sent back as output argument(s). Thus, one can create a function in the editor as shown:

$$function\ Avg = CalcStudentAvg\big(StdMrk, Wght\big)$$
$$Avg = StdMrk * Wght;$$

This function should now be saved in the current directory, and the m-file should be named after the function name: *CalcStudentAvg*. How do we then execute the function?

>> Average= *CalcStudentAvg*(StudentsMark, CourseWeighting);
Average =
13.6667
13.6667
15.3333

MATLAB provides a wide range of ways to perform computational tasks, and it is relatively easy to program it. This appendix does not intend to introduce all the possibilities, which will be impossible to do in a few pages, but the aim was to encourage interested readers to learn more about the various functionalities of MATLAB.

NOTE

1. These MATLAB files are available at https://www.crcpress.com/9780367207557.

Appendix A.2: Complex Numbers

The numbers that we regularly work with are called real numbers. Any number from minus infinity to plus infinity that we can imagine is a real number. Numbers like -10.333, -2.5, 0, 1, 2, 3.55 and 10 powered by 1000 are all real numbers. There is a one-to-one correspondence between the points on a line and real numbers. Thus, to represent real numbers geometrically, we would just need to show them on a line. However, to describe some physical phenomena, real numbers are not adequate for quantification purposes. One example of such a physical phenomenon is the frequency representation of signals. We should be able to describe the magnitude and phase of the Fourier transform together. Here is where the concept of complex numbers comes handy and can resolve many of our computational problems.

The complex numbers have two parts, namely, a real and an imaginary part. To geometrically represent these numbers, a single axis is not adequate—they should be presented on a plane. In other words, these numbers have two dimensions. Figure A.2.1 shows the geometrical representation of a complex number. Mathematically, a complex number (c) is written with its real part (a) and imaginary part (b) as $c = a + i.b$. One may then ask what i is in this notation, and that is the imaginary unit. However, there is an interesting relationship between the imaginary unit and the real numbers. The i satisfies $i^2 = -1$ or sometimes written as $i = \sqrt{-1}$. This is why the i is called the "imaginary" unit. No real number satisfies this relationship, as the square of any real number is always non-negative. Very often, j can also be used to denote the imaginary unit. MATLAB recognizes both i and j as the imaginary unit, provided you have not already defined them as variables named i and j. To avoid confusion, newer versions of MATLAB introduced $1i$ and $1j$ to denote the imaginary unit.

In the notation of a complex number $c = a + i.b$, a is the real part of c, and it is written as $Re(c) = a$, and $i.b$ is the imaginary part of c: $Im(c) = i.b$. One can perform arithmetic operations with complex numbers, such as addition, subtraction and multiplication.

If $c_1 = a_1 + i.b_1$ and $c_2 = a_2 + i.b_2$, then,

$$c_1 + c_2 = \left(a_1 + a_2\right) + i.\left(b_1 + b_2\right)$$

Subtraction is defined similarly. For multiplication, one can write:

$$c_1.c_2 = \left(a_1 a_2 - b_1 b_2\right) + i.\left(a_1 b_2 + a_2 b_1\right)$$

Note that in deriving the multiplication relationship, $i^2 = -1$ has been used.

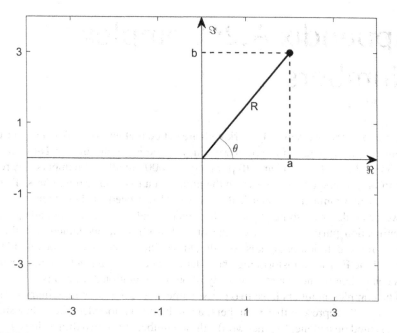

FIGURE A.2.1 The geometrical representation of a complex number with a real part (a) and imaginary part (b). The phase θ and the magnitude R of the complex number is used for polar representation of the complex number

The complex numbers can also be represented in what is known as "polar" representation. Instead of expressing a complex number c in terms of its real and imaginary parts, one can represent it in terms of its magnitude (R) and phase (θ).

$$c = Re^{i\theta}$$

Where $e^{i\theta} = cos(\theta) + i.sin(\theta)$, which is known as the Euler formula. Based on trigonometric equations, it is very straightforward to show that for a complex number $c = a + i.b$

$$R = \sqrt{a^2 + b^2} \text{ and } \theta = atan\left(\frac{b}{a}\right);$$

This is how the real and imaginary parts of a complex number are related to its magnitude and phase.

For a complex number like $c = a + i.b$, a complex-conjugate number is defined as $c^* = a - i.b$. Here, c^* is the complex-conjugate (or just conjugate) of c. The conjugate has some nice properties—for example, the multiplication of c and c^* equals the magnitude of c squared. In the polar representation $c^* = Re^{-i\theta}$, the conjugate has the same magnitude, but its phase has the opposite sign to the complex number c.

Appendix A.3: An Introduction to Convolution

A convolution is a mathematical operation that is very useful to describe a linear time-invariant (LTI) system response (see Section 7.2). Here, I do not explain how the mathematician ended up formulating the convolution operation but simply show how it looks mathematically and how it is implemented. I start with continuous time signals and then briefly introduce the discrete time version. Suppose a signal $x(t)$ is the input to an LTI system with an impulse response of $h(t)$ (see Section 7.2). One could also imagine $h(t)$ as another signal independent of its associated system; then $y(t)$ is defined as the convolution of $x(t)$ and $h(t)$ and one can write:

$$y(t) = x(t) * h(t) = \int_{-\infty}^{\infty} x(\tau)h(t-\tau)d\tau$$

For some specific signals, this integral can be analytically solved and $y(t)$ can be written as a function of time, but we can also numerically calculate this integral. If I just explain what this integral means in plain words, I would say that $h(\tau)$ needs to be flipped in time, meaning that one should find the mirror image of $h(\tau)$ with respect to the Y-axis and then shift the mirror image in time. One may now ask what this τ is? And that is what we know as the integral variable. Instead of trying to explain what that is in words, I will explain what this means in practice.

One should shift the mirror image to the left side until there is no overlap between the shifted mirror images of $h(\tau)$ and $x(\tau)$. Note that when there is no overlap between them, the integrand term $x(\tau)h(t-\tau)$ will be zero and the integral will simply be zero as well. Now if one shifts the mirror image to the right, there will be some overlap between the mirror images of $h(\tau)$ and $x(\tau)$, and the integrand is not zero any longer and one can calculate the integral. Figure A.3.1 illustrates this procedure for a hypothetical $x(t)$, which equals 1 within a $[-1,1]$ s time interval and an $h(t)$, which equals 1.1 within a $[0,0.5]$ s interval. The shaded area in the figure shows the overlap between the shifted mirror images of $h(\tau)$ and $x(\tau)$, and the dotted curve shows the result of calculating the convolution.

Also note that the convolution is commutative, meaning that

$$x(t) * h(t) = h(t) * x(t)$$

A similar operation can be done for discrete time signals, but the integration should now be converted to summation; therefore, for discrete time signals, the convolution between $x(n)$ and $h(n)$ is defined as

$$y(n) = x(n) * h(n) = \sum_{k=-\infty}^{\infty} x(k)h(n-k)$$

93

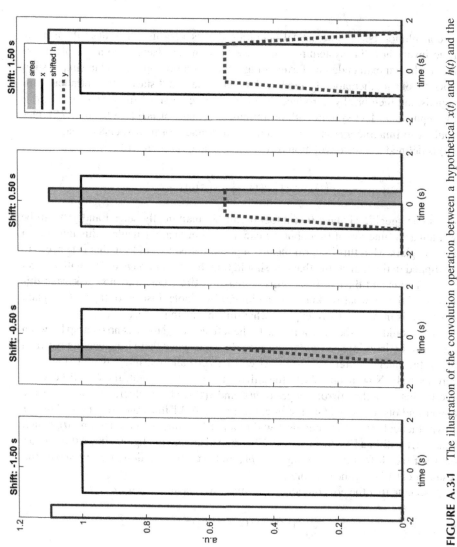

FIGURE A.3.1 The illustration of the convolution operation between a hypothetical $x(t)$ and $h(t)$ and the resulting $y(t)$ as the mirror image of $h(t)$ shifted in time

Appendix A.4: Correlation

Correlation in statistics is used to find out how two variables are associated with each other—for example, if one variable has an increasing trend, would the other one do the same, or would it be completely the opposite, or are we uncertain if the other variable is equally likely to take any of the two courses? In this area, several concepts are inter-related. I will introduce them very briefly. Without a loss of generality, I assume that I am working with signals whose means equal zero. Non-zero means only result in an offset in the calculations.

CORRELATION COEFFICIENT

The correlation coefficient (CC) is used to determine the degree of association between two signals. This can vary between −1 and 1. If CC is positive, this means that the two signals are positively correlated. An increasing/decreasing trend of one is likely associated with a similar trend in the other one. If CC equals 1, this association is perfect and any increasing/decreasing trend in one is associated with the same trend in the other one. If CC is negative, the two signals are anti-correlated, meaning that an increasing/decreasing trend of one is likely associated with an opposite trend in the other one. If CC equals −1, the two signals are perfectly anti-correlated, meaning that any increasing/decreasing trend in one of them is associated with an opposite trend in the other one. If CC equals 0, the two signals are uncorrelated, meaning that an increasing/decreasing trend of one is equally likely to be associated with the same or opposite pattern. In statistics, different versions of CC are defined, but here what I mean by CC is what is known as the Pearson correlation in statistics.

COVARIANCE

CC is closely related to another quantity known as covariance. As mentioned, CC is tightly related to the way two signals co-vary together, and that is what the covariance quantifies.

For two signals with zero mean, the covariance can be calculated as follows:

$$Covariance = \int_{-\infty}^{\infty} x(\tau)y(\tau)d\tau$$

For a discrete time signal, a summation should replace the integral. If the covariance of two signals is normalized by the squared root of the product of two signal variances, the results would be the CC.

CORRELATION FUNCTION

If one calculates the correlation between one of the signals and the shifted version of the other signal in time, we obtain a function of time (shifts,) which is known as correlation function. Mathematically, this can be denoted as (Oppenheim, Willsky, and Nawab, 1997) (page 65):

$$z(t) = \int_{-\infty}^{\infty} x(\tau) y(t + \tau) d\tau$$

In this case, I assume that the variance of $x(t)$ and $y(t)$ is normalized. This formulation is quite like the convolution formulation in Appendix A.3, and one can readily conclude that

$$z(t) = x(t) * y(-t)$$

If one needs it to be within [−1,1], the resulting integral should be divided by the square root of the product of signal variances. If the correlation function is calculated for a signal and its own shifted versions in time, it is called an auto-correlation function.

REFERENCE

A. V Oppenheim, A. S. Willsky, and S. H. Nawab, *Signals and Systems* (2nd ed.). NJ: Prentice Hall, 1997.

Index

9781032337814